Connecting Time and Space

Edited by Harry E. Bates

Selected Reprints

Published by
American Association of Physics Teachers

Connecting Time and Space
©1992 American Association of Physics Teachers

Published by:

American Association of Physics Teachers
5112 Berwyn Road
College Park, MD 20740-4100
U.S.A.

ISBN#: 0-917853-47-4

Contents

Preface v
H.E. Bates

**Resource Letter RMSL-1: Recent Measurements of the
Speed of Light and the Redefinition of the Meter** 1
H.E. Bates

Velocity of Light from the Molecular Constants of Carbon Monoxide ... 7
E.K. Plyler, L.R. Blaine, and W.S. Connor

Measurement of an Optical Frequency and the Speed of Light 13
Z. Bay, G.G. Luther, and J.A. White

**Characteristics of Tungsten-Nickel Point Contact Diodes
Used as Laser Harmonic-Generator Mixers** 17
E. Sakuma and K.M. Evenson

**Speed of Light from Direct Frequency and Wavelength
Measurements of the Methane-Stabilized Laser** 23
K.M. Evenson, J.S. Wells, F.R. Petersen, B.L. Danielson,
G.W. Day, R.L. Barger, and J.L. Hall

**Accurate Frequencies of Molecular Transitions
Used in Laser Stabilization: the 3.39-μm Transition in CH_4
and the 9.33- and 10.18-μm Transitions in CO_2** 27
K.M. Evenson, J.S. Wells, F.R. Petersen, B.L. Danielson, and G.W. Day

**Wavelength of the 3.39-μm Laser-Saturated
Absorption Line of Methane** 31
R.L. Barger and J.L. Hall

Locking a Laser Frequency to the Time Standard 35
Z. Bay and G.G. Luther

**Description, Performance, and Wavelengths
of Iodine Stabilized Lasers** 37
W.G. Schweitzer, Jr., E.G. Kessler, Jr., R.D. Deslattes, H.P. Layer, and J.R. Whetstone

**Determination of the Speed of Light by Absolute Wavelength
Measurement of the R[14] Line of the CO_2 9.4-μm Band
and the Known Frequency of This Line** 49
J.P. Monchalin, M.J. Kelly, J.E. Thomas, N.A. Kurnit, A. Szöke,
A. Javan, F. Zernike, and P.H. Lee

Measurement of the Speed of Light I. Introduction and Frequency Measurement of a Carbon Dioxide Laser 53
T.G. Blaney, C.C. Bradley, G.J. Edwards, B.W. Jolliffe, D.J.E. Knight,
W.R.C. Rowley, K.C. Shotton, and P.T. Woods

Measurement of the Speed of Light II. Wavelength Measurements and Conclusion 83
T.G. Blaney, C.C. Bradley, G.J. Edwards, B.W. Jolliffe, D.J.E. Knight,
W.R.C. Rowley, K.C. Shotton, and P.T. Woods

Precise Frequency Measurements in Submillimeter and Infrared Region 109
Y.S. Domnin, N.B. Kosheljaevsky, V.M. Tatarenkov, and P.S. Shumjatsky

Accurate Absolute Frequency Measurements on Stabilized CO_2 and He-Ne Infrared Lasers 113
A. Clairon, B. Dahmani, and J. Rutman

Improved Laser Test of the Isotropy of Space 119
A. Brillet and J.L. Hall

A Velocity of Light Measurement Using a Laser Beam 123
D.S. Edmonds, Jr., and R.V. Smith

Another Velocity of Light Experiment 127
B.G. Eaton, P.A. Johnson, and N.J. Petit

Measuring the Speed of Light by Independent Frequency and Wavelength Determination 129
H.E. Bates

A Pulser Circuit for Measuring the Speed of Light 135
M.E. Ciholas and P.M. Wilt

Preface

This reprint book is intended to provide the reader with an overview of events leading from the first crude measurements of the speed of light through more recent sophisticated laser-based measurements to the realization by the scientific community that separate length and time standards are not necessary. Since October of 1983, the speed of light has been defined and for metrological purposes the measurement of the speed of light can no longer be made. Any attempt to measure this quantity simply becomes an exercise in calibrating one's meterstick. The frequency-stabilized laser can now conveniently be used to interferometrically realize the meter in calibration setups. Time and space are now irrevocably connected in a practical way.

The series of papers selected for inclusion in this book not only tell the story of the redefinition of the meter but also emphasize a number of interesting technological developments that have made this a practical possibility. These include the development of precision interferometry, doppler-free laser frequency stabilization, the MIM diode mixer, and the measurement of optical frequencies.

The editor would like to acknowledge the contribution of Zoltan Bay who recommended the inclusion of a paper titled "Locking a Laser to a Time Standard" that was not included in the original resource letter. It represented a significant development in the steps required to finally define the speed of light and so it is included here. Special thanks to Roger H. Stuewer, Resource Letter Editor for the *American Journal of Physics*; Donna Willis, AAPT Director of Publications, Sean Elkin, Editorial Assistant, and Rebecca Rados, Graphic Designer. Many others have contributed to this effort. Please see acknowledgments in the resource letter.

Special thanks to Kenneth M. Evenson for allowing his photo to appear on the cover together with his apparatus for the measurement of laser frequencies.

Harry E. Bates
Towson State University
Towson, MD 21204
September 1992

Reprinted with permission from *American Journal of Physics* 56, 682–687, ©1988 American Association of Physics Teachers.

RESOURCE LETTER

Roger H. Stuewer, *Editor*
School of Physics and Astronomy, 116 Church Street
University of Minnesota, Minneapolis, Minnesota 55455

This is one of a series of Resource Letters on different topics intended to guide college physicists, astronomers, and other scientists to some of the literature and other teaching aids that may help improve course contents in specified fields. No Resource Letter is meant to be exhaustive and complete; in time there may be more than one letter on some of the main subjects of interest. Comments on these materials as well as suggestions for future topics will be welcomed. Please send such communications to Professor Roger H. Stuewer, Editor, AAPT Resource Letters, School of Physics and Astronomy, 116 Church Street SE, University of Minnesota, Minneapolis, MN 55455.

Resource Letter RMSL-1: Recent measurements of the speed of light and the redefinition of the meter

Harry E. Bates
Department of Physics, Towson State University, Towson, Maryland 21204

(Received 18 February 1988; accepted for publication 7 March 1988)

This Resource Letter provides a guide to the literature on recent measurements of the speed of light and the redefinition of the meter. The letter E after an item indicates elementary level or material of general interest to persons becoming informed in the field. The letter I, for intermediate level, indicates material of somewhat more specialized nature; and the letter A indicates rather specialized or advanced material. An asterisk (*) indicates those articles to be included in an accompanying Reprint Book.

I. INTRODUCTION

On 20 October 1983, the International Committee on Weights and Measurements agreed to a redefinition of the meter that fixes the value of the speed of light in vacuum to be 299 792 458 m/s. The scientific community has already experienced an effect of this new method of realizing the standard of length through a readjustment of the fundamental constants. See Table I. This landmark agreement was the product of a fascinating series of events in modern technology that linked the development of the laser with modern radio communication's technology.

Two graphs summarize selected measurements by a number of researchers whose papers are cited in this Resource Letter to provide the reader with a long-term trend in measurements. Specific citations to values in these figures are given in Table II. Figure 1 is a scatter plot of a number of values of the speed of light reported since 1927. The first value represents the culmination of Michelson's work during his Mount Wilson measurements. Note that the overall change in the measured value since then to the present is only about 0.008%. Figure 2 is a graph of the uncertainty in measured values plotted logarithmically against time. This Resource Letter provides references to papers playing a central role in the International Committee's definition, as well as the latest measurements of the speed of light prior to this definition. Some papers report student experiments to measure the speed of light. Also, a section of this paper deals with the question of the constancy of the speed of light and the isotropy of space.

II. HISTORICAL

The following selection of articles begins with an excellent historical account of the first observations of the finite speed of light made by Olaus Roemer and then goes on to describe some recent techniques for measuring the speed of light from 1949 to the present. An article referring to Michelson's 1927 measurements is also included. Basically the articles divide into phase-speed and group-speed measurements. The phase and group speeds would be the same in a vacuum or a nondispersive medium; however, they are different in air, and this systematic error must be corrected.

A. The first observations

An appropriate historical backdrop to this Resource Letter is provided by this detailed account of the events and misinterpretations surrounding Olaus Roemer's discovery.

1. "**de Mora Luminis: A Spectacle in Two Acts with a Prologue and an Epilogue,**" A. Wroblewski, Am. J. Phys. **53**, 620–630 (1985). (E)

B. Early phase-speed measurements

Phase-speed measurements obtained by observing the resonant frequencies of microwave cavities or by direct frequency counting were the most precise ones developed prior to laser measurements.

2. "**Velocity of Light and Radio Waves,**" L. Essen, Nature **165**, 582–583 (1950). (E)

Table I. Summary of the 1986 recommended values of the fundamental physical constants taken from Ref. 68. This abbreviated list of the fundamental constants of physics and chemistry is based on a least-squares adjustment with 17 degrees of freedom. The digits in parentheses are the 1-s.d. uncertainty in the last digits of the given value. Since the uncertainties of many of these entries are correlated, the full covariance matrix must be used in evaluating the uncertainties of quantities computed from them.

Quantity	Value	Units	Relative uncertainty (ppm)
Speed of light in vacuum	299 792 458	$m\,s^{-1}$	(exact)
Permeability of vacuum	$4\pi \times 10^{-7}$	NA^{-2}	
	= 12.566 370 614...	$10^{-7}\,NA^{-2}$	(exact)
Permittivity of vacuum	= 8.854 187 817...	$10^{-12}\,Fm^{-1}$	(exact)
Newtonian constant of gravitation	6.672 59(85)	$10^{-11}\,m^3 kg^{-1}\,s^{-2}$	128
Planck's constant	6.626 075 5(40)	10^{-34} J s	0.60
Elementary charge	1.602 177 33(49)	10^{-19} C	0.30
Magnetic flux quantum	2.067 834 61(61)	10^{-15} Wb	0.30
Electron mass	9.109 389 7(54)	10^{-31} kg	0.59
Proton mass	1.672 623 1(10)	10^{-27} kg	0.59
Proton–electron mass ratio	1836 152 701(37)		0.020
Fine-structure constant	7.297 353 08(33)	10^{-3}	0.045
Rydberg constant	10 973 731.534(13)	m^{-1}	0.0012
Avogadro constant	6.022 136 7(36)	$10^{23}\,mol^{-1}$	0.59
Faraday constant	96 485.309(29)	$C\,mol^{-1}$	0.30
Molar gas constant	8.314 510(70)	$J\,mol^{-1}\,K^{-1}$	8.4
Boltzmann constant	1.380 658(12)	$10^{-23}\,J\,K^{-1}$	8.5
Stefan–Boltzmann constant	5.670 51(19)	$10^{-8}\,W\,m^{-2}\,K^{-4}$	34
Non-SI units used with SI			
Electron volt	1.602 177 33(49)	10^{-19} J	0.30
(Unified) atomic mass unit	1.660 540 2(10)	10^{-27} kg	0.59

3. "The Velocity of Propagation of Electromagnetic Waves Derived From the Resonant Frequencies of a Cylindrical Cavity Resonator," L. Essen, Proc. R. Soc. London Ser. A **204**, 260–277 (1950). (A)
4. "A New Determination of the Free-Space Velocity of Electromagnetic Waves," K. D. Froome, Proc. R. Soc. London Ser. A **247**, 109–122 (1958). (A) See also Ref. 83.

C. Group-speed measurements

5. "Final Velocity-of-Light Measurements of Michelson," R. S. Shankland, Am. J. Phys. **35**, 1095–1096 (1967). (E) This article discusses the quality of Michelson's measurements and suggests that the Mount Wilson result was superior to the last result published under his name.

Table II. Selected list of speed of light measurements since 1927 with citations to references in this letter (see Figs. 1 and 2).

Ref. no.	Pub. date Mo./year	Experimenters	Speed of light km/s	Uncertainty km/s
5	12/27	Michelson at Mt. Wilson	299 798.9	4
82	1/35	Michelson et al. at Santa Ana	299 774	1.1
6	2/49	Bergstrand, geodimeter	299 796	2
2	4/50	Essen, NPL radar measurements	299 792	2.4
7	3/50	Bergstrand, geodimeter	299 792.700	0.25
3	6/50	Essen, NPL rf cavity	299 792.500	3.0
8	2/51	Bergstrand, geodimeter	299 793.100	0.2
9	10/52	Rank et al., spectral method	299 776	6
11	2/55	Plyler, spectral method	299 792	6
10	11/55	Rank et al., spectral method	299 791.9	2.2
4	3/58	Froome, millimeter wave	299 792.5	0.1
12	7/72	Bay et al., diff. freq. method	299 792.462	0.018
18	2/73	Evanson et al., direct freq.	299 792.4562	0.0011
32	2/73	Baird et al., wavelength meas.	299 792.458	0.002
34	9/74	Blaney et al., direct freq. meas.	299 792.459	0.0008
41	1/77	Blaney et al., direct freq. meas.	299 792.459	0.0006
27	7/77	Monchalin et al., direct freq. meas.	299 792.4576	0.0022
31	4/78	Woods et al., direct freq. meas.	299 792.4588	0.0002
54	1/87	Jennings et al., direct freq. meas. visible light	299 792.4586	0.0003

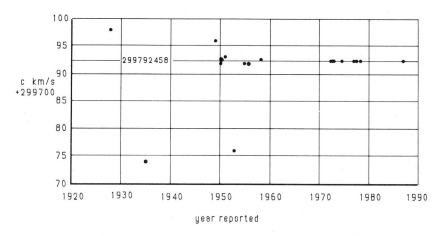

Fig. 1. A number of selected measurements of the speed of light since 1927. Values must be added to 299 700 km/s (from Table II).

The following group-speed measurements measure the envelope (or group) speed of light using the geodimeter. This device was based on the Kerr cell electro-optical modulator and was used to make a null measurement of the relative phase of the radio-frequency amplitude modulation between light beams traveling over long and short optical paths. See Ref. 82 for a description of this device.

6. "Velocity of Light and Measurement of Distances by High-Frequency Light Signalling," E. Bergstrand, Nature **163**, 338 (1949). (I)
7. "Velocity of Light," E. Bergstrand, Nature **165**, 405 (1950). (E)
8. "A Check Determination of the Velocity of Light," E. Bergstrand, Ark. Fys. **3**, 479–490 (1952). (I)

D. Indirect determinations of the speed of light

Optical rotational energy levels in molecules absorb and emit microwave radiation. The frequency of this radiation could be directly compared to frequency standards circa 1950. However, the wavelengths of molecular vibrational transitions are short enough for accurate measurements. If two vibrational transitions from a common upper state terminate on adjacent rotational levels in the lower state, a relationship between vibrational transitions (wavelength measurements) and rotational transitions (frequency measurements) exists from which the speed of light can be deduced.

9. **"Precision Determination of the Velocity of Light Derived From a Band Spectrum Method,"** D. H. Rank, R. P. Ruth, and K. L. Vander Sluis, J. Opt. Soc. Am. **42**, 693–698 (1952). (A)
10. **"Improved Value of the Velocity of Light Derived from a Band Spectrum Method,"** D. H. Rank, H. E. Bennett, and J. M. Bennett, Phys. Rev. **100**, 993 (1955). (A).
*11. **Velocity of Light from the Molecular Constants of Carbon Monoxide,"** E. K. Plyler, L. R. Blaine, and W. S. Connor, J. Opt. Soc. Am. **45**, 102–106 (1955). (A)

E. The application of the laser to speed of light measurements

In 1961 a conference on quantum electronics was held at Berkeley that provided a forum for discussing applications of the laser. Charles Townes predicted that ultrastable laser oscillators would make it possible to measure directly optical frequencies and that this would lead to very precise measurements of the speed of light. He observed that:

"...Maser techniques will probably make possible frequency multiplication from the radio frequency or microwave range into the visible and ultraviolet regions. It should be remembered that physicists have never directly measured the frequencies of infrared or optical radiation. Rather they measure wavelengths in this region, and then compute frequency from a knowledge of the velocity of light accurate to about one part in 10^6. Frequency multiplication up to the

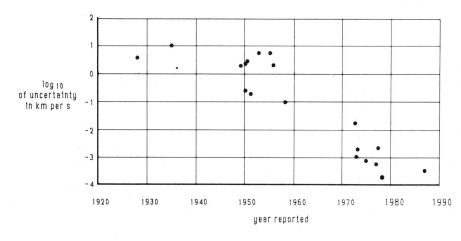

Fig. 2. A plot of the logarithm to the base 10 of the uncertainty in values of the speed of light as a function of the time the measurement was reported (from Table II).

visible region should allow these frequencies to be directly counted and hence measured in terms of our standard of time. Our standard of length is now defined in terms of a wavelength of visible light. Hence a measurement of its frequncy should allow immediately a determination of the velocity of light to a precision as great as that to which length is defined.

"A measurement of frequencies in the optical region to a precision greater than about one part in 10^8, to which length is now defined, will in fact connect length and time together throughout [sic] the velocity of light so firmly that perhaps separate standards of length and time would no longer be appropriate. Time, which can be measured and defined more precisely than can length, might be taken as the fundamental defined unit and length then derived from it by taking the velocity of light multiplied by some time interval as the standard of length..." [Second International Conference on Quantum Electronics, Berkeley, California, 1961. *Advances in Quantum Electronics*, edited by J. R. Singer (Columbia U.P., New York, 1961), pp. 4–11.]

1. The first laser measurement using differential interferometry

*12. "Measurement of an Optical Frequency and the Speed of Light," Z. Bay, G. G. Luther, and J. A.White, Phys. Rev. Lett. **29**, 189–192 (1972). (A) This paper discusses an ingenious technique that *seems* to have been made obsolete by the work cited in following sections.

2. Direct measurement of optical frequencies

The familiar ideas of frequency multiplication and superheterodyne detection of radio signals were first applied to frequency synthesis with light by Javan and his associates at MIT (Ref. 16). They used special point-contact diodes to measure directly the frequency of an HCN laser as compared to a harmonic of a 70-GHz klystron microwave generator. This technology grew out of work done on methods of detecting and generating harmonics of millimeter and submillimeter waves based on cat's-whisker-type detectors. In such a detector, a fine tungsten wire with a sharp point is brought into contact with a semiconductor crystal. The wire acts as a tiny antenna inducing currents in the nonlinear rectifying contact. Very large current densities are produced and large nonlinear responses observed. This trend toward bridging the gap between microwave and the optical regime started with an unusual observation reported by J. Dees in Ref. 13. Dees and his assistant, R. A. Miesch, who were then at the Martin Marietta Electromagnetics Laboratory in Orlando, Florida, discovered that a tungsten cat's whisker, when run into a metal post with no semiconducting crystal in place (a very thin oxide layer takes the place of the crystal), could produce a rectifying contact at very high frequencies. Although this detector was not as good as the silicon-crystal type in the microwave region, its extremely low capacitance gives it a very wide bandwidth and thus it turns out to be useful at optical frequencies. This eventually permitted the measurements of frequencies of visible light and led to the redefinition of the meter. It is interesting to note that the principle of the metal–insulator–metal (MIM) diode was explained much earlier (see Ref. 81).

13. "Detection and Harmonic Generation in the Submillimeter Wavelength Region," J. W. Dees, Microwave J. **9**, 48–55 (1966). (A)

14. "Extension of Laser Harmonic-Frequency Mixing Techniques into the 9 μm Region With an Infrared Metal–Metal Point-Contact Diode," V. Daneu, D. Skoloff, A. Sanchez, and A. Javan, Appl. Phys. Lett. **15**, 398–401 (1969). (I).

15. "Heterodyne Detection of 10.6-μm Radiation by Metal-to-Metal Point Contact Diodes," R. L. Abrams and W. B. Gandrud, Appl. Phys. Lett. **17**, 150–152 (1970). (I)

16. Fundamental and Applied Laser Physics Proceedings of the Esfahan Symposium 29 August to 5 September 1971, edited by M. S. Feld, A. Javan, and N. Kumit (Wiley, New York, 1973), pp. 295–334. (A)

*17. "Characteristics of Tungsten–Nickel Point Contact Diodes Used as Laser Harmonic-Generator Mixers," E. Sakuma and K. M. Evenson, IEEE J. Quantum Electron. **QE-10**, 599–603 (1974). (A)

3. The first measurements of the speed of light using direct optical frequency synthesis

*18. "Speed of Light from Direct Frequency and Wavelength Measurements of the Methane-Stabilized Laser," K. M. Evenson, J. S.Wells, F. R. Petersen, B. L. Danielson, G. W. Day, R. L. Barger, and J. L. Hall, Phys. Rev. Lett. **29**, 1346–1349 (1972). (A)

*19. "Accurate Frequencies of Molecular Transitions Used in Laser Stabilization: The 3.39 μm Transition in CH_4 and the 9.33 μm and 10.18 μm Transitions in CO_2," K. M. Evenson, J. S. Wells, F. R. Petersen, B. L. Danielson, and G. W. Day, Appl. Phys. Lett. **22**, 192–195 (1973). (A)

*20. "Wavelength of the 3.39 μm Laser-Saturated Absorption Line of Methane," R. L. Barger and J. L. Hall, Appl. Phys. Lett. **22**, 196–199 (1973). (A)

F. Redefinition of the meter

On 20 October 1983, The International Committee on Weights and Measurements agreed to a redefinition of the meter by defining the speed of light to be 299 792 458 m/s. The most accurate measurements of length are now based on the wavelength of light from stabilized lasers whose frequencies have been measured.

21. "Proposed New Definition of the Meter," David T. Goldman, J. Opt. Soc. Am. **70**, 1640–1641 (1980). (E)

22. "News from the BIPM," P. Giacomo, Metrologia **20**, 25–30 (1984). (E) This article summarizes the 17th Conference Generale des Poids et Mesures (CGPM) including the following decisions: (a) The meter is the length of path traveled by light in vacuum during a time interval of 1/(299 792 458) of a second; and (b) The definition of the meter in force since 1960 based upon the transition between the levels $2p_{10}$ and $5d_5$ of the atom of Krypton 86 is abrogated. These decisions were approved 20 October 1983.

23. "Documents Concerning the New Definition of the Metre," Metrologia **19**, 163–177 (1984). (I) The official collection of references and notes on the redefinition of the meter. This article points out that this is only the second redefinition of the meter since the unit length was defined in terms of the international prototype meter by the first CGPM in 1889.

24. "The New Definition of the Meter," P. Giacomo, Am. J. Phys. **52**, 607–613 (1984). (I)

II. STABILIZATION OF LASERS

Soon after the invention of the laser, investigators realized the potential of this device as a frequency standard because of its narrow output linewidth. However, since the operating frequency is determined by cavity length together with the absolute frequency and profile of the gain line, early lasers could not be utilized as frequency standards. Techniques for locking the frequency of lasers to well-defined features of atomic or molecular transitions thus proved to be a necessary step toward the redefinition of the meter.

25. **Laser Spectroscopy,** F. R. Petersen, D. G. McDonald, J. D. Cupp, and B. L. Danielson, Proceedings of the Vail, Colorado Conference, edited by R. G. Brewer and A. Mooradian (Plenum, New York, 1973), pp. 555–569. (A)

*26. "Description, Performance, and Wavelengths of Iodine Stabilized Lasers," W. G. Schweitzer, Jr., E. G. Kessler, Jr., R. Deslattes, H. P. Layer, and J. R. Whetstone, Appl. Opt. **12**, 2927–2938 (1973). (A)

*27. "Determination of the Speed of Light by Absolute Wavelength Measurement of the R(14) Line of the CO_2 9.4 μm Band and the Known Frequency of This Line," J. P. Monchalin, M. J. Kelly, J. E. Thomas, N. A. Kurnit, A. Szoke, A. Javan, F. Zernike, and P. H. Lee, Opt. Lett. **1**, 5–7 (1977). (A) See also Ref. 42.

28. "Phase-Lock Control Considerations for Coherently Combined Lasers," J. B. Armor, Jr. and S. Robinson, Jr., Appl. Opt. **18**, 3165–3175 (1979). (A)

29. "Frequency-Modulation Spectroscopy: A New Method for Measuring Weak Absorption and Dispersions," G. C. Bjorklund, Opt. Lett. **5**, 15–17 (1980). (I)

30. "Investigation of the Stability of the Emission Wavelength of a Laser with an External Neon Absorption Cell," V. P. Kapralov, V. E. Privalov, and E. G. Chulyaeva, Sov. J. Quantum Electron. **10**, 1062–1064 (1980). (I)

III. LASER MEASUREMENTS

31. "Frequency Determination of Visible Laser Light by Interferometric Comparison With Upconverted CO_2 Laser Radiation," P. T. Woods, B. W. Jolliffe, W. R. C. Rowley, K. C. Shotton, and A. J. Wallard, Appl. Opt. **17**, 1048–1054 (1978). (I)

32. "Wavelength of the CH_4 Line at 3.39 μm," K. M. Baird, D. S. Smith, and W. E. Berger, Opt. Commun. **7**, 107–109 (1973). (A)

33. "Absolute Frequency Measurement of the R(12) Transition of CO_2 at 9.3 Microns," T. G. Blaney, C. C. Bradley, G. J. Edwards, J. J. E. Knight, P. T. Woods, and B. W. Jolliffe, Nature **244**, 504 (1973). (A)

34. "Measurement of the Speed of Light," T. G. Blaney, C. C. Bradley, G. J. Edwards, B. W. Jolliffe, D. J. E. Knight, W. R. C. Rowley, K. C. Shotton, and P. T. Woods, Nature **251**, 46 (1974). (A)

35. "Accurate Wavelength Measurement on Upconverted CO_2 Laser Radiation," B. W. Jolliffe, W. R. C. Rowley, K. C. Shotton, A. J. Wallard, and P. T. Woods, Nature **251**, 46–47 (1974). (A)

36. "Determination of the Absolute Oscillation Frequency of the $2P_4$ Transition in a Helium–Neon Laser," V. M. Geller and G. I. Grif, Sov. J. Quantum Electron. **4**, 1052–1053 (1975). (A)

37. "Laser Wavelength Comparison by High Resolution Interferometry," H. P. Layer, R. D. Deslattes, and W. G. Schweitzer, Jr., Appl. Opt. **15**, 734–743 (1976). (A)

38. "Absolute Frequencies of the Methane Stabilized He–Ne Laser (3.39 microns)," T. G. Blaney, G. J. Edwards, B. W. Jolliffe, D. J. E. Knight, and P. T. Woods, J. Phys. D: Appl. Phys. **9**, 1323–1330 (1976). (I)

39. "Frequency of the Methane Stabilized He–Ne Laser at 3.39 Microns Measured Relative to the 10.17 Micron R(32) Transition of the CO_2 Laser," B. G. Whitford and D. S. Smith, Opt. Commun. **20**, 280–283 (1977). (I)

*40. "Measurement of the Speed of Light I. Introduction and Frequency Measurement of a Carbon Dioxide Laser," T. G. Blaney, C. C. Bradley, G. J. Edwards, B. W. Jolliffe, D. J. E. Knight, W. R. C. Rowley, K. C. Shotton, and P. T. Woods, Proc. R. Soc. London Ser. A **355**, 61–88 (1977). (A)

*41. "Measurement of the Speed of Light II. Wavelength Measurements and Conclusion," T. G. Blaney, C. C. Bradley, G. J. Edwards, B. W. Jolliffe, D. J. E. Knight, W. R. C. Rowley, K. C. Shotton, and P. T. Woods, Proc. R. Soc. London Ser. A **355**, 89–114 (1977). (A)

42. "Determination of the Speed of Light by Absolute Wavelength Measurement of the R(14) Line of the CO_2 9.4 μm Band and the Known Frequency of This Line: Errata," J. P. Monchalin, M. J. Kelly, J. E. Thomas, N. A. Kurnit, A. Szoke, A. Javan, F. Zernike, and P. H. Lee, Opt. Lett. **1**, 140 (1977). (A) See also Ref. 27.

43. "Frequency Measurements in the Optical Range: Current Status and Prospects," V. M. Klement'ev, Y. A. Matyugin, and V. P. Chebotaev, Sov. J. Quantum Electron. **8**, 953–958 (1978). (A)

44. "Confirmation of the Currently Accepted Value 299 792 458 Metres per Second for the Speed of Light," K. M. Baird, D. S. Smith, and B. G. Whitford, Opt. Commun. **31**, 367–368 (1979). (I)

45. "Comment on 'Confirmation of the Currently Accepted Value 299 792 458 Metres per Second for the Speed of Light, by K. M. Baird, D. S. Smith, and B. C. Whitford," W. R. C. Rowley, K. C. Shotton, and P. T. Woods, Opt. Commun. **34**, 429–430 (1980). (I)

46. "Wavelength of a Helium–Neon Laser Stabilized by Saturated Absorption in Iodine at 612 nm," S. J. Bennett, P. Cerez, J. Hamon, and A. Chartier, Metrologia **15**, 125 (1979). (I)

47. "Frequency of the Methane-Stabilized He–Ne Laser at 88 THZ Measured to +3 Parts in 10^{11} (For New Metre Definition)," D. J. E. Knight, G. J. Edwards, P. R. Pearce, and N. R. Cross, Nature **285**, 388–390 (1980). (I)

48. "Measurement of the Frequency of the 3.39 Micron Methane Stabilized He–Ne Laser," D. J. E. Knight, G. J. Edwards, P. R. Pearce, and N. R. Cross, IEEE Trans. Instrum. Meas. **IM-29**, 257–263 (1980). (A)

*49. "Precise Frequency Measurements in Submillimeter and Infrared Region," Y. S. Domnin, N. B. Kosheljaevsky, V. M. Tatarenkov, and P. S. Shumjatsky, IEEE Trans. Instrum. Meas. **IM-29**, 264–267 (1980). (A)

*50. "Accurate Absolute Frequency Measurements on Stabilized CO_2 and He–Ne Infrared Lasers," A. Clairon, B. Dahmani, and J. Rutman, IEEE Trans. Instrum. Meas. **IM-29**, 268–272 (1980). (A)

51. "Measurement of the Frequency of a He–Ne/CH_4 Laser," Y. S. Komnin, N. B. Koshelyaevskii, V. M. Tatarenkov, and P. S. Shumyatskii, JETP Lett. **34**, 167–170 (1981). (I)

52. "Accurate Laser Wavelength Measurement with a Precision Two-Beam Scanning Michelson Interferometer," J. P. Monchalin, M. J. Kelly, J. E. Thomas, N. A. Kurnit, A. Szoke, F. Zernike, P. H. Lee, and A. Javan, Appl. Opt. **20**, 736–757 (1981). (A)

53. "Laser Frequency Measurements and the Redefinition of the Meter," P. Giacomo, IEEE Trans. Instrum. Meas. (USA) **IM-32**, 244–246 (1983). (E)

54. "The Continuity of the Meter: The Redefinition of the Meter and the Speed of Visible Light," D. A. Jennings, R. E. Drullinger, K. M. Evenson, C. R. Pollock, and J. S. Wells, J. Res. NBS **92**, 11–16 (1987). (I)

IV. INVESTIGATIONS OF THE ISOTROPY OF SPACE

55. "The Rod Contraction-Clock Retardation Ether Theory and the Special Theory of Relativity," H. Erlichson, Am. J. Phys. **41**, 1068–1077 (1973). (A)

*56. "Improved Laser Test of the Isotropy of Space," A. Brillet and J. L. Hall, Phys. Rev. Lett. **42**, 549–552 (1979). (A)

57. "A Suggestion to Detect the Anisotropic Effect of the One-Way Velocity of Light," T. Chang, J. Phys. A: Math. Gen. **13**, L207–L209 (1980). (A)

58. "On Random Fluctuations of the Velocity of Light in Vacuum," B. N. Belyaev, Sov. Phys. J. (USA) **23** (11), 942–946 (1980). (A)

59. "Fact and Illusion in the Speed-of-Light Determinations of the Romer Type," L. Karlov, Am. J. Phys. **49**, 64–66 (1981). (I) See Ref. 67.

60. "On the Impossibility of Measuring the One-Way Velocity of Light by Means of the Stellar Aberration," T. Sjodin, M. F. Podlaha, Lett. Nuovo Cimento **31**, 433–436 (1981). (A)

61. "Laser Interferometry Experiments on Light-Speed Anisotropy," H. Aspden, Phys. Lett. A (Netherlands) **85A**, 411–414 (1981). (I)

62. "Can an Anisotropic Effect of the One-Way Velocity of Light Really be Measured?," A. Flidrzynski and A. Nowicki, J. Phys. A: Math. Gen. **15** (3), 1051–1052 (1982). (A)

63. "Can One Measure the One-Way Velocity of Light?," C. Nissim-Sabat, Am. J. Phys. **50**, 533–536 (1982). (A)

64. "Mirror Reflection Effects in Light Speed Anisotropy Tests," H. Aspden, Speculations Sci. Technol. (Switzerland) **5** (4), 421–431 (1982). (I)

65. "Proposed Method of Measuring First-Order Speed of Light Anisotropy," H. Aspden, Phys. Lett. A (Netherlands) **92A**, 165–166 (1982). (I)

66. "The Special Theory of Relativity and the One-Way Speed of Light," B. Townsend, Am. J. Phys. **51**, 1092–1096 (1983). (A)

67. "Note on the Laboratory Romer Method for Determining the Speed of Light," L. Karlov, Am. J. Phys. **52**, 873 (1984). (I) See Ref. 59.

V. IMPACT OF DEFINITION OF THE SPEED OF LIGHT ON THE ADJUSTMENT OF PHYSICAL CONSTANTS

CODATA is an interdisciplinary committee of the International Council of Scientific Unions which deals with data of importance to people working in the area of science and technology. Partly as a result of the new definition of the meter in terms of a defined speed of light, a new report has been published readjusting the fundamental physical constants. This report is the result of a five-year effort by the committee. Table I lists new adjusted values of the fundamental physical constants given in the reference below. It's worth noting that the permittivity of the vacuum is now exact as a result of the redefinition of the meter.

68. "The 1986 Adjustment of the Fundamental Physical Constants," E. R. Cohen and B. N. Taylor, CODBA (**63**), 1–32 (1986). (A)

VI. STUDENT LABORATORY MEASUREMENTS OF THE SPEED OF LIGHT

A few student laboratory measurements and demonstrations using a number of diverse techniques are included below for those interested in developing laboratories for undergraduates. Phase-velocity measurements allow the student to explore both frequency-measuring techniques using harmonic mixing and length-measuring techniques using geometric optics or interferometry. For group velocity measurements, timing using coincidence instrumentation or the oscilloscope must be employed.

69. "Speed of 'Light' Measurement," W. F. Huang, Am. J. Phys. **38**, 1159 (1970). (E)
*70. "A Velocity of Light Measurement Using a Laser Beam," D. S. Edmonds and R. V. Smith, Am. J. Phys. **39**, 1145–1148 (1971). (E) (Note this apparatus won a merit award in the 1971 AAPT apparatus competition.)
71. "Measuring the Speed of Light with a Laser and Pockels Cell," D. N. Page and C. D. Geilker, Am. J. Phys. **40**, 86–88 (1972). (E)
72. "Speed of Light Determined by Microwave Interference," L. Bergel and S. Arnold, Am. J. Phys. **44**, 546–547 (1976). (E)
73. "Determination of the Speed of Light by Measurement of the Beat Frequency of Internal Laser Modes," R. G. Brickner, L. A. Kappers, and F. P. Lipschultz, Am. J. Phys. **47**, 1086–1087 (1979). (I)
74. "Measurement of the Speed of Light Using Nuclear Timing Techniques," S. S. Sherbini, Am. J. Phys. **48**, 578–579 (1980). (E)
*75. "Another Velocity of Light Experiment," B. G. Eaton, P. A. Johnson, and N. J. Petit, Phys. Teach. **18**, 667–668 (1980). (E)
76. "Minimal Apparatus for the Speed-of-Light Measurement, R. E. Crandall, Am. J. Phys. **50**, 1157–1159 (1982). (E)
*77. "Measuring the Speed of Light by Independent Frequency and Wavelength Determination," H. E. Bates, Am. J. Phys. **51**, 1003–1008 (1983). (E)
78. "Simple First-Order Test of Special Relativity," J. Byl, M. Sanderse, and W. van der Kamp, Am. J. Phys. **53**, 43–45 (1985). (I) This paper is unique in this section in that it reports on a student test of the isotropy of the speed of light rather than measurement of the absolute speed of light.
79. "Estimating the Speed of Light with a TV Set," M. C. Schroeder and C. W. Smith, Phys. Teach. **23**, 360 (1985). (E)
*80. "A Pulser Circuit for Measuring the Speed of Light," M. E. Ciholas and P. M. Wilt, Am. J. Phys. **55**, 853–854 (1987).

VII. REVIEW ARTICLES, BOOKS, AND OTHER RELATED PAPERS

81. **Crystal Rectifiers**. H. C. Torrey and C. A. Whitmer (McGraw-Hill, New York, 1948) and (Boston Technical, Lexington, MA, 1964), 2nd ed. (A) Page 70 introduces the idea that a MIM diode could be useful at high frequencies.
82. **Velocity of Light**. J. H. Sanders (Pergamon, New York, 1965). (E–A) Introductory material and a reprint of reports of three important measurements of the speed of light by Michelson, Pease, and Pearson; Essen and Gordon-Smith; and Bergstrand.
83. **The Velocity of Light and Radio Waves**, K. D. Froome and L. Essen (Academic, New York, 1969). (E–A) An excellent summary of work up to 1969.
84. "Some Recent Determinations of the Velocity of Light III," J. F. Mulligan, Am. J. Phys. **44**, 960–969 (1976). (I) The last in a series of nice review articles in this Journal since 1952.
85. "Prospects for Precision Physical Experiments in Optics," E. V. Baklanov and V. P. Chebotaev, Sov. Phys. Usp. **20**, 631–637 (1977). (A)
86. "Stabilized Lasers and Precision Measurements," J. H. Hall, Science **202**, 147–156 (1978). (I) This article traces the development of stabilized lasers from the early work of Javan and others up to 1978 and including the use of lasers as applied to the measurement of the speed of light and the redefinition of the meter.
87. "Time, Frequency and Physical Measurement," H. Hellwig, K. M. Evenson, and D. J. Wineland, Phys. Today **31** (12), 23–30 (1978). (E)
88. "First Direct Frequency Measurement of Visible Light Reported," K. M. Evenson, D. A. Jennings, and F. R. Petersen, Dimensions NBS (USA) **63**, 20–22 (1979). (E)
89. "Laser Frequency Measurements and the Meter," K. M. Evenson, Laser Focus (USA) **17**, 61–63 (1981). (E)
90. "Foundations of the International System of Units (SI)," R. A. Nelson, Phys. Teach. **19**, 596–613 (1981). (E) This article traces the history of the International Bureau of Weights and Measures and the evolution of the International System of Units (SI).
91. "Speed of Light Outside the Solar System: A New Test Using Visual Binary Stars," R. Gruber, D. Koo, and J. Middleditch, Publ. Astron. Soc. Pac. (USA) **93**, 777–782 (1981–1982). (A)
92. **Quantum Metrology and Fundamental Physical Constants**, edited by P. H. Cutler and A. A. Lucas (Plenum, New York, 1983), p. 658. This is the proceedings of a NATO Advanced Study Institute on Quantum Metrology and Fundamental Constants, held 16–28 November 1981. The article, "Speed of Light, Historical Review to 1972" by K. M. Baird (pp. 165–179) (E) summarizes developments to that date. Another, "Frequency Measurements From the Microwave to the Visible, the Speed of Light, and the Redefinition of the Meter," by K. M. Evenson (pp. 181–207) (A) discusses modern work based on direct optical-frequency measurements.
93. "A Hypothetical Experiment. Measurement of the Velocity of Light from Electrical Measurements on a Loss-Free Dielectric," G. P. Owen, J. Opt. (India) **11**, 433–434 (1982).
94. "Frequency Measurement of Radiation," K. M. Baird, Phys. Today **36** (1), 52–57 (1983). (E)
95. "Laser Frequency Standards," S. N. Bagaev and V. P. Chebotaev, Sov. Phys. Usp. **29**, 82–103 (1986). (A)
96. "Navigation and the Speed of Light," A. A. Bartlett, Am. J. Phys. **54**, 9 (1986). (E)
97. "Roemer, Navigation, and the Speed of Light," A. J. Friedman, Am. J. Phys. **54**, 583 (1986). (E).

ACKNOWLEDGMENTS

The author would like to thank the following persons for their helpful comments and discussions during the preparation of this work: C. O. Alley, K. M. Baird, H. R. Crane, K. M. Evenson, J. J. Gallagher, D. L. Goodstein, J. L. Hall, R. W. McMillan, R. A. Nelson, A. L. Schawlow, J. R. Smithson, R. H. Stuewer, B. N. Taylor, and C. H. Townes. This work was supported in part by the Faculty Research Committee of Towson State University.

Velocity of Light from the Molecular Constants of Carbon Monoxide

EARLE K. PLYLER, L. R. BLAINE, AND W. S. CONNOR
National Bureau of Standards, Washington, D. C.
(Received October 26, 1954)

By infrared spectroscopy precise measurements of the wavelengths of forty-three rotational lines in the CO absorption band at 4.67 μ were made. Fifteen lines from $J=32$ to $J=52$ in the P branch were measured from the emission spectrum of CO. These data were reduced by the method of least squares, and values of the rotational constants, B_0 and D_0, of CO were calculated. A second, slightly less precise set of measurements over approximately the same range of J values was similarly reduced, and combined estimates of the constants were obtained. Using the resulting value of B_0 with the value obtained from microwave frequency measurements, the velocity of light was estimated to be 299 792 km/sec. The estimated standard deviation of this value is 6 km/sec.

INTRODUCTION

THE measurement of the velocity of light has been the subject of many investigations. In 1941 Birge[1] considered the results reported to that time and obtained a weighted average of 299 776±4 km/sec. Another value, 299 773±10 km/sec, has been deduced by Dorsey[1] in considering measurements up to 1944. More recently, several new determinations by different observers[2] have given an average value of 299 793 km/sec with an estimated error of less than 1 km/sec.

A recent method, which involves the measurement of molecular constants by microwave and infrared spectroscopy, was suggested by Dr. A. E. Douglas of Ottawa and was carried out by Rank, Ruth, and Van der Sluis.[3] The rotational lines are measured in microwave spectra in Mc/sec, and the frequencies of the lines are represented by the formula

$$\nu = 2B_0(J+1) - 4D_0(J+1)^3, \quad (1)$$

where B_0 is proportional to the reciprocal of the moment of inertia in the ground state, D_0 is the centrifugal stretching constant in the ground state, and J is the rotational quantum number. The rotational lines of a rotational-vibrational band are measured in the infrared region and the wave numbers are represented by the formula

$$\nu = \nu_0 + (B_1+B_0)m + (B_1-B_0-D_1+D_0)m^2 \\ -2(D_1+D_0)m^3 - (D_1-D_0)m^4, \quad (2)$$

where B_0 and D_0 are the same as in Eq. (1), B_1 and D_1 are the same type of constants for the upper state, the band center is designated ν_0 and $m=J+1$ for the R branch, and $m=-J$ for the P branch of the band. By using B_0 calculated from Eq. (1), measured in Mc/sec, with that of B_0 in Eq. (2), measured in cm^{-1}, we obtain the velocity of light. The relation is

$$c = B_{0\,\text{micro}} / B_{0\,\text{IR}}.$$

In 1952 Rank, Ruth, and Vander Sluis made measurements on the (004) and (103) infrared bands of HCN. By the use of a Fabry-Perot interferometer they were able to measure the spectral lines with high precision. After the data were reduced they obtained a value of the velocity of light of 299 776±6 km/sec. This 1952 value, agreeing with the older measurements, raised several questions: is it to be expected that B_0 determined from microwave spectra is different from B_0 determined from infrared spectra, and are Eqs. (1) and (2) adequate to represent precise measurements? The question of possible shifts of the positions of the lines with pressure has been raised. There is no experimental evidence for a shift. Measurements show that the lines are symmetrical within experimental error. On account of these questions a set of measurements was started on the molecular constants of carbon monoxide, to see what value this molecule would give for the velocity of light. Carbon monoxide was selected because it can be measured both in absorption and in emission. The large J values, present in flame spectra, are helpful in determining precisely the constants of Eq. (2).

After the experimental data were obtained on the fundamental band of carbon monoxide, two reports appeared which essentially answered the questions raised in the previous paragraph. In the first report Rank (see reference 9 on page 105) revised the earlier determination[3] to 299 776±19 km/sec. This value of

---

[1] R. T. Birge, Revs. Modern Phys. **13**, 233 (1941); Repts. Progr. in Phys. **8**, 92–101 (1941); N. E. Dorsey, Trans. Am. Phil. Soc. **34**, Part 1 (1944).
[2] J. W. DuMond and E. R. Cohen, Revs. Modern Phys. **25**, 691 (1953).
[3] Rank, Ruth and Van der Sluis, J. Opt. Soc. Am. **42**, 693 (1952).

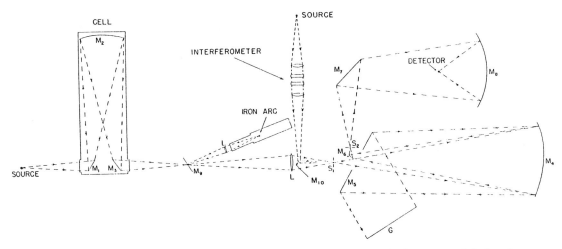

Fig. 1. A diagram showing the optical arrangement of the spectrometer, the standard sources, and the interferometer. The mirror M_9 is mounted on a lever and may be inserted or removed from the beam. A detailed description of the instrument and its operation is given in the text.

the velocity of light now overlaps the more recent determinations. The second article reported a new determination of the velocity of light by measurements of the (002) band of HCN. In this article Rank, Shearer, and Wiggins (see reference 8 on page 105) reported a value of 299 789.8±3 km/sec for the velocity of light. This last value is confirmed by the results for the velocity of light of 299 792±6 km/sec, which have been obtained in the measurements of carbon monoxide in the present work. Thus there are two experiments using infrared and microwave measurements, carried out in different laboratories, which report values of the velocity of light that overlap, and both values agree with the more recent determinations of the velocity of light by other methods.

EXPERIMENTAL METHOD

In order to measure the rotational-vibrational band of CO at 4.67 μ with high precision, the fringes of a Fabry-Perot interferometer were used for a comparison spectrum. The arrangement of the interferometer was similar to that used by Douglas and Sharma[4] and will not be described. A double pen recorder was used for obtaining the fringe system with one pen and the absorption of the carbon monoxide with the second pen. The fringes were of visible white light in the region from 5000 to 6000 A and they were detected in higher orders of the grating with a 1P28 photomultiplier and an accompanying amplifier. The carbon monoxide lines were detected with a cooled PbTe cell. The wave numbers of the fringes were determined in reference to the absorption lines by superimposing on the recorded fringe system higher orders of emission lines of xenon which were detected by the PbTe cell. A set of fringes formed with a 2-mm Invar separator with superimposed standard lines was used for measurements. The maxima

[4] A. E. Douglas and D. Sharma, J. Chem. Phys. **21**, 448 (1953).

of the fringe system were separated by about 0.3 cm^{-1} in the region of 4.5 μ. Emission lines of xenon were selected which fell beyond the two limits of the band and spectra were obtained of the standard lines, absorption lines of CO, and the fringes without interruption of the scanning of the spectrometer.

The errors of the grating, the uneven rate of driving, and temperature changes introduced variations in the apparent spacing of the fringe system. In fact, the fringe system was used to determine the amount of these variations for the spectrometer. It was found that the distance on the chart between adjacent fringes varied by about 2 percent. The variation was random and by inspection was of such magnitude as to produce an average error in the absorption lines of CO of ±0.002 cm^{-1}. The average variation in the measured lines in absorption as judged by the differences between the calculated and measured lines in Table I was about ±0.005 cm^{-1}. This indicates that the total error in measurement was about 2.5 times as large as the instrumental error. Thus the total error in measurement arises primarily in determining the centers of the absorption lines.

The optical arrangement of the grating spectrometer, the absorption cell, and of the sources is shown in Fig. 1. A modified form of the Pfund system is used for the optical arrangement. The radiation falling on S_1 passes through an aperture in the plane mirror M_5 and falls on the paraboloidal mirror M_4, which is figured on axis. The radiation is returned in a parallel beam to M_5 from which it passes to the grating and returns to M_5 and M_4 which brings it to a focus on S_2. By the use of the mirrors M_7 and M_8, the radiation is focused on the detector. This instrument was originally designed for the long wavelength region and optics of large aperture were used. M_4 is 35 cm in diameter and 70 cm in focal length. The instrument was found to give good resolution in the 3 to 6-μ region with 8-in.

gratings having 7500 lines/in. and 4600 lines/in. Sharp lines separated by 0.15 cm^{-1} can be partially resolved by the instrument. The mirror M_9 is on a lever so that it can be introduced into the beam or removed. Sources for standard lines, such as the iron vacuum arc and krypton and neon lamps, have their radiation focused on the mirror M_9. The small mirror M_{10} is used in placing the interferometer beam on the slit, and it is placed either outside the cone of radiation from the lithium fluoride lens, or in the section of the beam corresponding to the opening in plane mirror M_5. The absorption cell has a total optical path of one meter and the carbon monoxide spectrum was obtained at pressures of about one cm Hg.

RESULTS

Two sets of observations were made and the wave numbers of the rotational lines were determined. With low pressures of CO, the lines were sufficiently intense to permit measurements up to $J=23$ in the R branch, and to $J=28$ in the P branch. By the use of a torch burning methane-oxygen, we measured the emission lines of CO from $J=32$ to $J=52$ in the P branch. Whenever a rotational line was overlapped by another CO line from the isotopic bands or higher states in emission, it was omitted. In Table I are listed the observed and calculated positions of the lines for the more precise set of observations. The difference between the calculated and observed positions of lines are also listed in the last column of Table I. The lines measured in absorption vary from the calculated values as obtained by the least squares reduction by about ±0.005 cm^{-1}; in emission the variation is about ±0.01 cm^{-1}. The data were reduced by least squares, and the constants of Eq. (2) were obtained. The calculations were made both with the m^4 term and without it. The absolute value of the constant D_1-D_0 from the more precise set of measurements was found to be 2.99×10^{-9} cm^{-1}. Since only a small number of the rotational lines were determined for $J>30$ it was felt that the data were not adequate to permit a precise determination to be made of this constant. The absolute value of D_1-D_0 from theory is 1.8×10^{-9} cm^{-1}. Goldberg[5] has measured some CO lines in the harmonic band from the sun with values of J to 68, and has obtained 1.95×10^{-9} cm^{-1} for the absolute value of D_1-D_0.

TABLE I. Table of observed and calculated rotational lines of CO at 4.7 μ.

R	Observed cm^{-1}	Calculated cm^{-1}	Experimental-calculated	R	Observed cm^{-1}	Calculated cm^{-1}	Experimental-calculated
R 23	2224.694	2224.690	0.004	14	2086.322	2086.322	0.000
22	2221.732	2221.728	0.004	15	2082.009	2082.002	0.007
21	2218.733	2218.728	0.005	16	2077.650	2077.649	0.001
20	2215.685	2215.689	−0.004	17	...	2073.265	
19	2212.600	2212.612	−0.012	18	2068.851	2068.845	0.006
18	2209.498	2209.497	0.001	19	...	2064.397	
17	2206.345	2206.343	0.002	20	2059.911	2059.912	−0.001
16	2203.147	2203.153	−0.006	21	2055.391	2055.397	−0.006
15	2199.929	2199.924	0.005	22	...	2050.851	
14	2196.661	2196.658	0.003	23	2046.271	2046.272	−0.001
13	2193.357	2193.355	0.002	24	2041.663	2041.661	0.002
12	2190.010	2190.014	−0.004	25	2037.030	2037.020	0.010
11	2186.636	2186.637	−0.001	26	2032.349	2032.347	0.002
10	2183.226	2183.223	0.003	27	2027.635	2027.642	−0.007
9	2179.761	2179.772	−0.011	28	2022.899	2022.908	−0.009
8	2176.287	2176.284	0.003	29	...	2018.153	
7	2172.759	2172.760	−0.001	30	...	2013.340	
6	...	2169.200		31	...	2008.517	
5	2165.602	2165.603	−0.001	32	2003.659	2003.660	−0.001
4	...	2161.971		33	1998.767	1998.772	−0.005
3	2158.309	2158.303	0.006	34	1993.867	1993.854	0.013
2	2154.596	2154.599	−0.003	35	1988.910	1988.906	0.004
1	...	2150.860		36	1983.919	1983.929	−0.010
R 0	...	2147.085		37	1978.923	1978.921	0.002
P 1	2139.432	2139.430	0.002	38	1973.871	1973.884	−0.013
2	2135.554	2135.550	0.004	39	1968.805	1968.818	−0.013
3	2131.639	2131.635	0.004	40	...	1963.720	
4	...	2127.685		41	...	1958.596	
5	2123.700	2123.702	−0.002	42	1953.459	1953.444	0.014
6	2119.677	2119.684	−0.007	43	1948.252	1948.262	−0.010
7	...	2115.632		44	...	1943.051	
8	2111.555	2111.546	0.009	45	1937.831	1937.812	0.019
9	2107.413	2107.425	−0.012	46	1932.555	1932.544	0.001
10	2103.265	2103.272	−0.007	47	1927.246	1927.248	−0.002
11	2099.096	2099.084	0.012	48	1921.926	1921.925	0.001
12	2094.870	2094.863	0.007	49	...	1916.573	
13	2090.603	2090.609	−0.006	50	...	1911.186	
				51	1905.776	1905.786	−0.010

[5] Leo Goldberg and Edith A. Muller, Astrophys. J. **118**, 397 (1953).

TABLE II. Constants of CO.

	1st set of observations in cm^{-1}	2nd set of observations in cm^{-1}	Combined constants in cm^{-1}
$\nu_0=$	2143.272 ± 0.003	2143.275 ± 0.002	2143.274 ± 0.0015
$B_1=$	1.90501 ± 0.00007	1.90498 ± 0.00004	1.904994 ± 0.000036
$B_0=$	1.922528 ± 0.000070	1.922521 ± 0.000043	1.922523 ± 0.000037
$D_0=$	$6.26 \pm 0.042 \times 10^{-6}$	$6.25 \pm 0.03 \times 10^{-6}$	$6.26 \pm 0.025 \times 10^{-6}$

The estimated values of the constants given below are averages of the estimates from the two sets of observations. Except for the velocity of light, the attached plus and minus values are estimated standard deviations. For the velocity of light the attached value is the estimated standard deviation plus an allowance for a possible absolute error.

If an estimate of a constant were normally distributed and if the standard deviation of the estimate were known exactly, then the probability that confidence limits determined in this way (i.e., by adding the standard deviation to, and subtracting it from, the estimate) would enclose the true constant is 68 percent. If twice the standard deviation were added and subtracted, then the associated probability would be 96 percent. These assumptions are so closely realized that the probabilities quoted here are approximately correct.

The error in the value of D_1-D_0 from our results is too small to have any importance in the determination of D_0. A value of 6.26 ± 0.025 cm^{-1} is found for D_0. In some recent microwave measurements of Bedard, Gallagher, and Johnson[6] on the second rotational line of CO, the value of D_0 was found to be 6.30 ± 0.03 cm^{-1}. These two values are in excellent agreement for two different types of measurements. By use of the previous measurements of Gilliam, Johnson, and Gordy[7] on the first rotational line of CO, both B_0 and D_0 are known from microwave spectra. By using the two microwave measurements we found B_0 to be 57 635.65 Mc/sec.

The other constants of Eq. (2) are: $\nu_0 = 2143.274 \pm 0.0016$ cm^{-1}, $B_1 = 1.904\,994 \pm 0.000036$ cm^{-1}, and $B_0 = 1.922\,523 \pm 0.000037$ cm^{-1}. In Table II are given the values of the constants for the two sets of data. The speed of light is determined with a confidence limit as

$$C = 299\,792 \pm 6 \text{ km/sec.}$$

This value is obtained by using the quotient B_0 (microwave) to B_0 (infrared). The absolute value of ν_0 is believed to be accurate to better than ± 0.01 cm^{-1}. Included in the error of ± 6 km/sec is the estimated standard deviation of C, which is 5.7 km/sec, and an allowance of ± 0.5 km/sec for the possible displacement of ν_0.

The measurements of the frequencies of the rotational lines in the CO fundamental band were made in the radiometry section and the reduction of the data by least squares was carried out in the mathematics division. The method of the reduction of data is given in the last section of this paper. (The calculations were also confirmed by the Bureau's electronic computer.)

The questions raised in the introduction in regard to the first determination by Rank, Ruth, and Van der Sluis are answered now that the velocity of light measured by the microwave-infrared method is in agreement with the recent values found by other methods. The most recent value obtained by Rank, Shearer, and Wiggins[8] on the (002) band of HCN is 299 789.8±3 km/sec and this value for the velocity of light is in the same range as found in this work from measurements on CO. Rank[9] has recently revised his first determination of the velocity of light to 299 776±19 km/sec on the basis of estimates of systematic errors which he attributed to his grating. This revised limit of confidence of the measurements of Rank, Ruth, and Van der Sluis now overlaps the more recent determinations of the velocity of light by other methods.

In conclusion it is now clear from measurements in two laboratories that the molecular constant method yields values which are in accord with other recent measurements. A good summary of these earlier values is given by DuMond and Cohen.[2] With the experience gained in these measurements it is evident that future measurements would lead to a smaller measurement error, and that a large number of trials on one band would reduce the standard deviation. This procedure, however, would not take into consideration any systematic errors which were not properly evaluated.

THE STATISTICAL ANALYSIS OF THE DATA

The polynomial

$$\nu = \nu_0 + (B_1 + B_0)m + (B_1 - B_0)m^2 - 4Dm^3$$

was fitted to each set of data by the method of least squares. In this section the details of the fittings and the methods for combining the results will be explained.

Consider either of the sets of data. Suppose that it contains N observations, which may be denoted by $y_1, y_2, \cdots y_N$. If the error in an observation is denoted

---

[6] Bedard, Gallagher, and Johnson, Phys. Rev. **92**, 1440 (1953).
[7] Gilliam, Johnson, and Gordy, Phys. Rev. **78**, 140 (1950).
[8] Rank, Shearer, and Wiggins, Phys. Rev. **94**, 575 (1954).
[9] D. H. Rank, J. Opt. Soc. Am. **44**, 341 (1954).

by e, then the ith observation may be represented as

$$y_i = \nu_0 + (B_1 + B_0) m_i + (B_1 - B_0) m_i^2 - 4D m_i^3 + e_i \quad (i = 1, \cdots, N).$$

If the e's are independent, have true average zero, and belong to the same population of errors, then the method of least squares[10] requires that the estimates of ν_0, $B_1 + B_0$, $B_1 - B_0$, and $-4D$ be such that the sum of the squares of the differences between the observed and calculated y's be minimized. In the case at hand the e's do not appear to belong to the same population of errors since the lines in absorption appear to be measured more precisely than the lines in emission. However, this difference in precision is not great and has been disregarded.

By differentiating the expression

$$\sum_{i=1}^{N} (y_i - \nu_0 - (B_1 + B_0) m_i - (B_1 - B_0) m_i^2 + 4D m_i^3)^2$$

successively with respect to ν_0, $B_1 + B_0$, $B_1 - B_0$, and $-4D$ and setting the derivatives equal to zero, the following normal equations are obtained:

$$N[\nu_0] + \sum m[(B_1+B_0)] + \sum m^2[(B_1-B_0)] + \sum m^3[(-4D)] = \sum y,$$

$$\sum m[\nu_0] + \sum m^2[(B_1+B_0)] + \sum m^3[(B_1-B_0)] + \sum m^4[(-4D)] = \sum my,$$

$$\sum m^2[\nu_0] + \sum m^3[(B_1+B_0)] + \sum m^4[(B_1-B_0)] + \sum m^5[(-4D)] = \sum m^2 y,$$

$$\sum m^3[\nu_0] + \sum m^4[(B_1+B_0)] + \sum m^5[(B_1-B_0)] + \sum m^6[(-4D)] = \sum m^3 y.$$

Brackets have been put around the unknowns to denote estimates of them.

This system of equations may be solved for the estimates. The solutions are of the form

$$[\nu_0] = C_{11} \sum y + C_{12} \sum my + C_{13} \sum m^2 y + C_{14} \sum m^3 y,$$
$$[B+B] = C_{21} \sum y + C_{22} \sum my + C_{23} \sum m^2 y + C_{24} \sum m^3 y,$$
$$[B-B] = C_{31} \sum y + C_{32} \sum my + C_{33} \sum m^2 y + C_{34} \sum m^3 y,$$
$$[-4D] = C_{41} \sum y + C_{42} \sum my + C_{43} \sum m^2 y + C_{44} \sum m^3 y,$$

where the C's are functions of N, $\sum m$, $\cdots$ $\sum m^6$, and $C_{12} = C_{21}$, $\cdots$, $C_{34} = C_{43}$.

The C's in the latter system of equations are of importance in determining the precisions of the estimates. Before considering them, however, it is necessary to describe the measure of the precision of the observations themselves. Let the variance of the y's (i.e., the true average of the e's) be denoted by σ^2. Then the standard deviation σ may be used to measure the precision of the observations. If σ^2 is unknown, it is estimated from the formula

$$[\sigma^2] = \frac{1}{(N-4)} \sum_{i=1}^{N} [e_i^2],$$

where $[e_i] = y_i - [\nu_i]$, and $[\nu_i] = [\nu_0] + [(B_1+B_0)] m_i + [(B_1-B_0)] m_i^2 [-4D] m_i^3$.

The variance of $[\nu_0]$ is $C_{11} \sigma^2$,

of $[B_1] = \frac{1}{2}[(B_1+B_0)] + [(B_1-B_0)]$
is $\frac{1}{4}(C_{22} + C_{33} + 2C_{23}) \sigma^2$,

of $[B_0] = \frac{1}{2}[(B_1+B_0)] - [(B_1-B_0)]$
is $\frac{1}{4}(C_{22} + C_{33} - 2C_{23}) \sigma^2$,

and of $[D]$ is $\frac{1}{16} C_{44} \sigma^2$. The corresponding standard deviations are obtained by taking square roots. By adding (and subtracting) some multiple of the standard deviation to (from) an estimate, confidence limits may be determined for the estimate.

The estimates from the two sets may be efficiently combined by weighting according to the reciprocals of their estimated variances.[11]

---

[10] R. L. Anderson and T. A. Bancroft, *Statistical Theory in Research* (McGraw-Hill Book Company, Inc., New York, 1952).

[11] W. G. Cochran and S. P. Carroll, Biometrics **9**, 447–459 (1953).

Measurement of an Optical Frequency and the Speed of Light

Z. Bay, G. G. Luther, and J. A. White*

National Bureau of Standards, Washington, D. C. 20234
(Received 12 May 1972)

> We report the measurement of the frequency of the 633-nm red laser line. This is the first measurement of an optical frequency in the visible range without reference to the speed of light or to a measured wavelength. Combination of the optical frequency with the known wavelength yields c to an accuracy higher than previously known. This method demonstrates the practicability of a single-standard time-length measurement system unified via a defined value of the speed of light.

Progress in an experiment whose principle was described previously[1-3] has led recently to the determination of the optical frequency of the red (632.9-nm) He-Ne laser line and a redetermination of the speed of light. The method is based on the idea that if the difference and ratio of two optical frequencies are known, then the optical frequencies themselves are known.

By electro-optic modulation of the laser light (frequency ν) at the microwave frequency ω, the sideband frequencies $\nu \pm \omega$ are generated. The two sidebands are introduced into an evacuated Fabry-Perot cavity. The length L of the cavity and ω are simultaneously adjusted so that both sidebands pass the cavity with maximum intensity. When this is accomplished, the ratio of the two frequencies is the ratio of the two order numbers (N_+ and N_-, respectively) in the cavity. Then the optical frequency can be expressed as

$$\nu = [(N_+ + N_-)/(N_+ - N_-)]\omega \tag{1}$$

or

$$\nu = (N/n)2\omega, \tag{2}$$

where

$$N = (N_+ + N_-)/2 \tag{3}$$

is the order number of ν, and

$$n = N_+ - N_- \tag{4}$$

is the order number belonging to the beat note between the two sideband frequencies, 2ω. The interferometer thus establishes a relationship between the optical and the microwave frequencies.

Thus, the optical frequency is determined by measuring a microwave frequency, which is directly related to the primary frequency standard, and by determining the two order numbers N and n. Note that the measurement of ν in terms of ω makes reference only to the unit of time. It does not depend on any definition of a unit of length. It does not require knowledge of a wavelength, of the speed of light or of the length of the cavity. It merely depends upon tuning the cavity simultaneously to the two sidebands.

The condition of maximum transparency for both sidebands is produced by two servo loops as shown in Fig. 1. After passing the interferometer the sidebands are directed to two etalons, each of which passes only one sideband. The length of the interferometer is modulated at a frequency of approximately 40 kHz. The first-harmonic signals from the photomultipliers are added and phase detected in one circuit, subtracted and phase detected in another. The sum of the two first harmonics is used to servo L, via a piezoelectric transducer, to minimize that sum. The difference of the two first harmonics servos ω via a voltage-controlled oscillator to minimize that difference. (The frequency of this oscillator is added to a high-order multiple of the in-house 100-kHz frequency, derived from a frequency standard.) Clearly, if the sum and the difference of the two first harmonics are zero, the first-harmonic signals themselves are zero; i.e., the interferometer is tuned to both sidebands simultaneously and the conditions for Eqs. (1) and (2) are satisfied.

The precision of locking L to ν is similar in this experiment to that in other experiments in which a high-finesse interferometer is locked to a laser. The finesse of our interferometer is about 500, and a precision within 1% of the width of the transparency curve is easily achieved.

There are several favorable features of this experiment which allow the setting of ω to L with high precision. (i) Since the sidebands are generated by the microwave modulation of ν, the measurement of their separation, 2ω, on the frequency scale is limited only by the stability of the frequency standard. (ii) The two first har-

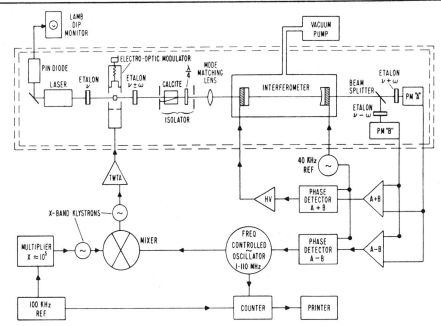

FIG. 1. Modulation interferometer for determining optical frequencies. A stable optical frequency ν is generated by the laser. Microwave frequencies $\pm\omega$ are added to ν in an electro-optic crystal mounted in the microwave cavity. The etalons shown are tuned to reject frequencies other than those indicated. The length L of the interferometer and frequency ω are adjusted, via servos, to achieve maximum transmission simultaneously at $\nu+\omega$ and $\nu-\omega$. The frequency ν is then related simply to ω (see text). Portions of the experiment included inside the dotted lines are contained in a mechanically and thermally isolated box.

monics belonging to the sidebands are strongly correlated. Thus their difference, which is used to drive ω, is insensitive to short-term fluctuations in ν, in L, or in the optics and the air path between the laser and the interferometer. (iii) To diminish uncorrelated noise between the sideband signals (e.g., shot noise) long time constants can be used in the servoing of ω. The "effective" time constant is finally given by the time of counting of ω. Successive counts, each taken for a counting time of 100 sec, fluctuated only by a few hundred hertz. This corresponds to about 10^{-4} of the bandwidth of the interferometer, or to a few parts in 10^8 for $\omega \sim 10^{10}$ Hz for the present experimental configuration.

Counts extended over 10–20 min showed slow variations in excess of noise fluctuations, amounting to a few kilohertz. We attribute these to optical feed-back effects and expect improvements by further refinements in the isolation between the Fabry-Perot and the laser. Observations showed these slow variations to be of random character. We decided, therefore, to use the equipment in its present state to produce, concurrently with other types of experiments being performed in the infrared, a preliminary measurement of ν (and hence of c) in the visible spectrum.

In the evaluation of the data, one has to take into account that N and n in Eq. (2) are not integers, even for maximum transmission of both sidebands by the interferometer. Their deviations from integral numbers, caused by the reflection phase shifts of the mirrors and by diffraction phase shifts, are treated by Bay and Luther,[3] and by Bay.[4]

To eliminate the reflection phase shifts, the interferometer is set successively to two different lengths, L_1 and L_2, and the corresponding ω_1 and ω_2 measured. Then[3,4]

$$\nu = \frac{N_2 - N_1 + \Phi_2 - \Phi_1}{n_2/2\omega_2 - n_1/2\omega_1}, \qquad (5)$$

where $N_2 - N_1$, n_2, and n_1 are now integral numbers, and Φ_2 and Φ_1 (the diffraction phase shifts) are calculated to better than 10^{-3} by cavity theory.[5] Diffraction phase shifts do not appear in the denominator of Eq. (5) because cavity theory shows that those phase shifts depend only on the geometry of the cavity and not on wavelength.

TABLE I. Estimated systematic errors.

100-kHz in-house frequency	$<10^{-9}$
Diffraction phase-shift correction	$<10^{-9}$
Phase uncertainties from mirror irregularities	$<10^{-9}$
Zeroing of electronic servoing	$<10^{-9}$
Wavelength of Lamb-dip laser (relevant for c measurement only)	$<1.5\times10^{-8}$

Consequently, the diffraction phase shifts for the two sideband waves are identical. The small errors due to irregularities of the mirror surfaces and slightly different mode diameters of the two sideband waves have been treated in Ref. 4. Their estimated values for the present experiment are included in Table I.

In the experiments $L_1 \sim 12$ cm, $L_2 \sim 25$ cm, $\omega_1 \sim \omega_2 \sim 1.034\times 10^{10}$ Hz were used. This results in the values $n_1 = 17$, $n_2 = 35$.

For the determination of $N_2 - N_1$ in Eq. (5) a trial number for ν is used. Since $N_2 \sim 8\times 10^5$, $N_1 \sim 4\times 10^5$, and $N_2 - N_1 \sim 4\times 10^5$, a trial number for ν with an uncertainty of less than 10^{-6} can be used to exclude the possibility of an error of one unit in N_2 and N_1. Such a number is available, since c is known to within 3×10^{-7},[6] and our λ is known to be $(632.99147\pm 1)\times 10^{-5}$ nm by measurements of C. Sidener at the National Bureau of Standards. It should be noted that the use of such a trial number does not in any way bias our measurement of ν or c. The measurements replace the trial number by the measured value for ν.

In the course of taking data, the length of the interferometer was alternated between the positions corresponding to $n_1 = 17$ and $n_2 = 35$. In each position, several counts of ω were taken, corresponding to different order numbers N_1 and N_2, respectively. Their departure from integral numbers, $\mathcal{E}_1$ and $\mathcal{E}_2$, respectively, were calculated and checked for statistical randomness. From 37 determinations of $\mathcal{E}_1$ and 36 determinations of $\mathcal{E}_2$, we obtained a correction to the trial number ν of -1.3×10^{-7} with the statistical standard deviation of the mean of $\pm 6\times 10^{-8}$. All known systematic errors, for which estimates are given in Table I, contribute substantially less than this value. This leads to the following conclusions:

(1) The frequency of the red laser line used is

$\nu = 473\,612\,166 \pm 29$ MHz.

This is the first absolute determination of an optical frequency in the visible spectrum. This frequency is about 5 times higher than the highest infrared frequency yet measured.[7] It should be noted that this method of measurement is applicable to any laser line, and that it relates the optical to the microwave frequency in one step, without the use of intermediary (infrared) frequencies. The method can be applied conveniently to tunable lasers.

(2) From the above measurement of ν and the measured wavelength, the speed of light is calculated to be

$c = 299\,792.462 \pm 0.018$ km/sec.

The accepted value of c $(299\,792.50\pm 0.1$ km/sec$)$[8] is the same as measured by Froome.[6] Our error bar is smaller by a factor of about 5 than that reported by Froome.

(3) These experiments are a prototype of length measurements which are based on c and the measurement of microwave frequencies.[4] Indeed, the denominator in Eq. (5), multiplied by c, gives $2(L_2 - L_1)$. Thus, lengths are determined to optical precision without the need to know an optical wavelength or an optical order number, if c is known to the desired accuracy or if the value of c is defined. Alternatively, the method can be used to make a most direct measurement of c in the present units of time and length by simultaneously determining $L_2 - L_1$ in terms of the Kr^{86} wavelength.

(4) These experiments demonstrate the feasibility of determining optical order numbers via modulation techniques.[4] Indeed, the product $\nu n/2\omega$ determines immediately the integral part of the optical order number. These techniques can replace fringe counting and optical multiplication methods which are especially difficult and inconvenient for long interferometers.

Future improvements in the accuracy of this method can be achieved by (a) diminishing the effects of optical feed back, (b) increasing the length of the interferometer, and (c) increasing the microwave frequency. Since (b) and (c) increase the precision linearly, there can be little doubt that, still confined to a table top, the experiment can yield two more orders in the precision.

These improvements can lead to a measurement of the speed of light limited in accuracy only by that of the Kr^{86} wavelength standard. For such a c measurement in the visible spectrum the 633-nm He-Ne laser locked to the Lamb dip in the present experiments can be replaced by one locked to the saturated absorption of the I$_2$

molecule.⁹

The ultimate accuracy of optical-frequency measurements and that of length measurements by this method is expected to be limited only by imperfections of mirror surfaces.⁴ This technological limitation is applicable also to wavelength comparisons and length measurements based on any wavelength standards, irrespective of their possible better quality. Thus, since they are applicable throughout the entire spectrum, these experiments demonstrate the possibility and practicability of a unified time-length measurement system[2,3,4,10,11] based on a frequency standard and on a defined value of the speed of light, compatible with the present meter but otherwise arbitrary.

We wish to acknowledge the contributions of several people at the National Bureau of Standards, especially R. G. Carpenter, C. Sidener, R. O. Stone, B. N. Taylor, and K. W. Yee.

---

*Permanent address: Department of Physics, The American University, Washington, D. C. 20016.

[1]Z. Bay and H. S. Boyne, in *Quantum Electronics and Coherent Light, International School of Physics "Enrico Fermi," Course XXXI, 1963*, edited by C. H. Townes and P. A. Miles (Academic, New York, 1964), p. 352.

[2]Z. Bay and G. G. Luther, Appl. Phys. Lett. 13, 303 (1968).

[3]Z. Bay and G. G. Luther, in *Precision Measurements and Fundamental Constants*, National Bureau of Standards Special Publication No. 343, edited by D. N. Langenberger and B. N. Taylor (U. S. GPO, Washington, D. C., 1971), p. 63.

[4]Z. Bay, in *Precision Measurements and Fundamental Constants*, National Bureau of Standards Special Publication No. 343, edited by D. N. Langenberger and B. N. Taylor (U. S. GPO, Washington, D. C., 1971), p. 59.

[5]H. Kogelnik and T. Li, Proc. IEEE 54, 1312 (1966).

[6]K. D. Froome, Proc. Roy. Soc., Ser. A 247, 109 (1958).

[7]K. M. Evenson, G. W. Day, J. S. Wells, and L. O. Mullen, Appl. Phys. Lett. 20, 133 (1972).

[8]B. N. Taylor and W. H. Parker, and D. N. Langenberg, Rev. Mod. Phys. 41, 375 (1969).

[9]G. R. Hanes and K. M. Baird, Metrologia 5, 32 (1969); J. D. Knox and Yoh-Han Pao, Appl. Phys. Lett. 8, 360 (1971).

[10]Z. Bay, in *Proceedings of the Fourth International Conference on Atomic Masses and Fundamental Constants, Teddington, England, September 1971* (Plenum, New York, 1972).

[11]Z. Bay and J. A. White, Phys. Rev. D 5, 796 (1972).

Characteristics of Tungsten–Nickel Point Contact Diodes Used as Laser Harmonic-Generator Mixers

EIICHI SAKUMA AND KENNETH M. EVENSON

Abstract—Properties of tungsten–nickel point contact diodes, when used as harmonic-generator mixers, were measured. The measured properties are those which will be useful to workers wishing to use these high-speed devices (faster than 10^{-14} s). Some of the properties measured were: the decrease in signal with mixing order; the detected signal as a function of laser frequency; and the power required to optimize the harmonic beat notes.

INTRODUCTION

RECENTLY, the cesium beam frequency standard and the krypton wavelength standard were linked in a definitive speed of light experiment by measuring both the frequency and wavelength of a 3.39-μm (88-THz) methane-stabilized He–Ne laser [1]–[3]. In extending frequency measurements to this high value, tungsten on nickel (W–Ni) point contact diodes were used as harmonic-generator mixers. These room temperature diodes were first reported in 1966 in a submillimeter wave application [4] and then used at laser frequencies in 1969 [5]. The extension of their use to 88 THz by the generation of 88 THz from the third harmonic of 29 THz in 1972 [6] demonstrated a response to at least 1.1×10^{-14} s. In a more recent experiment, sideband reradiation from the diode has been reported at 30 THz [7].

There are at least four different theories of operation for the point contact diode: 1) tunneling through an oxide layer [8]–[12]; 2) field emission [13]; 3) a quantum mechanical-scattering explanation [14]; and 4) a geometrically induced asymmetrical tunneling [15]. We wish to report the measurement of some characteristics of these diodes some of which differ from those recently reported

Manuscript received February 25, 1974; revised April 22, 1974.
E. Sakuma was with the Quantum Electronics Division, Institute for Basic Standards, National Bureau of Standards, Boulder, Colo. 80302. He is now with the National Research Laboratory of Metrology, Itabashi-Ku, Tokyo, Japan.
K. M. Evenson is with the Quantum Electronics Division, Institute for Basic Standards, National Bureau of Standards, Boulder, Colo. 80302.

[16]. It is hoped that these data will be useful both to those working toward a physical explanation of the operation of the diode as well as to those just wishing to build and use these devices as harmonic-generator mixers.

The diodes have been chiefly used to synthesize (and hence measure) extremely high frequencies (laser frequencies), and we concentrated on measurements to characterize their use as harmonic-generator mixers. The RF beat note signal intensity itself as generated in the diode was measured as a function of mixing order, driving power, and bias voltage. DC impedances, sensitivity as a detector, and the optimum driving power were also observed.

To measure an unknown frequency ν_u, a radio frequency difference (a beat note) between ν_u and a frequency synthesized from harmonics of several sources ($l\nu_l \pm m\nu_m \pm n\nu_n$) is generated by a mixing process in the same diode in which the harmonic generation occurred. That is, the diodes are irradiated with all of the frequencies and the beat note ν_{RF} developed across the diode is given by

$$\nu_{RF} = \nu_u - (l\nu_l \pm m\nu_m \pm n\nu_n)$$

where ν_u is a laser frequency, ν_l and ν_m are lower basis or reference laser frequencies and ν_n is a microwave frequency. The quantity within the parentheses is the synthesized frequency close to the unknown laser frequency ν_u. l, m, and n are harmonic numbers and the mixing order is $(1 + |l| + |m| + |n|)$. For example, $\nu_{H_2O} = 12 \nu_{HCN} + \nu_n$, where $\nu_{H_2O} = 10.718$ THz, $\nu_{HCN} = 0.891$ THz, $\nu_n = 29$ GHz, $l = 12$, $m = 0$, and $n = 1$.

The Measurements

Open structure diodes were employed which were identical to those used previously [1], [6]. The optically polished end of a 99.5-percent-pure nickel wire 2 mm in diameter served as the base element of the diode. When other samples of nickel with purities from 98 to 99.99 percent were tried, there was no correlation between purity and signal intensity. Beat note strengths from sample to sample varied by as much as 20 dB and the 99.5-percent samples exhibited somewhat smaller variability. A 25-μm-diameter tungsten wire (the type typically used in field emission devices) electrochemically etched to a point less than 2000 Å in radius served as the point element and antenna. The tungsten side of the diode was connected to a 0-100-MHz amplifier with a 3-dB noise figure: its output was connected to a conventional spectrum analyzer.

The diodes were somewhat finicky and delicate and changed during use; however, several different contacts could be made before repointing the tungsten catwhisker. A necessary prerequisite for obtaining a "good" beat note (and the "good" signals varied by many decibels) was to observe a rectified voltage which responded as a square wave at a chopping frequency of 500-1000 Hz. To use the tungsten-nickel point contact diode, the drive signal (the oscillator generating the harmonic near the unknown frequency) should have sufficient power to generate a millivolt or more of rectified voltage on the diode (usually 1-100 mW of power and careful focusing is required). Powers from the unknown oscillator and the other oscillators are adjusted so the rectified voltage on the djode is about 1/4-1/10th of that from the drive oscillator; at these levels, and after trying several diodes, a beat signal is generally detectable.

The largest beat signals were obtained when the tungsten whisker was driven negative by the applied radiation. It should be mentioned that beat notes could be obtained when the rectified signal was of the opposite polarity; however, the signals were much weaker under these circumstances.

To find the response of the diodes as a function of the mixing order, microwaves as well as lasers were used because of a lack of sufficient laser line sources. Three microwave sources were used which supplied 50-100 mW of power. Two were stabilized X- and K-band klystrons which supplied power to irradiate the diodes from open waveguides located about 2-3 cm from the whisker antenna. The third was a variable-frequency unstabilized (1.7-4.1-GHz) commercial microwave oscillator which was coupled capacitively with a 4-cm-long copper wire connected to the end of a coaxial cable and located approximately 1 cm from the whisker. The power coupled to the diode was not measured but we estimate that it was only a few percent of the power available.

To test the operation of the diode at laser frequencies, linearly polarized radiation from HCN (311 and 337 μm), H_2O (28 μm), and CO_2 (9.3 μm) lasers was used. Available maximum powers were 75, 200, and 500 mW, respectively. Coupling of laser radiations to the 2-mm-long tungsten antenna is explained by the application of well-known antenna theory at these wavelengths [6], [17]. S/N ratios of the RF beat notes at various mixing orders were measured with the signal optimized at each one by varying the oscillator power.

As the power was increased, the beat signal and noise level varied such that an optimum S/N ratio could be achieved. In all cases, one could obtain a maximum S/N by adjusting the power of each source independently. Fig. 1 shows the general behavior of the optimized RF beat notes versus total mixing order. These data were taken using many diodes, and each point is an averaged value. The RF beat notes were between 20 and 40 MHz. The third-order point at 73 dB is questionable since it was later discovered that the spectrum analyzer saturated above 70 dB. The results of mixing two microwave frequencies are shown by the circles. Here, $\nu_u \cong l\nu_l$ with $\nu_u = 8.1$ GHz, and $l = 2, 3$, and 4, or with $\nu_u \cong 19.94$ GHz and $l = 5, 6, 7, 8, 9$, and 10. The results of mixing three microwave frequencies is shown by the squares where $\nu_u \cong \nu_l + m\nu_m$, with $\nu_u = 19.94$ GHz, $\nu_l = 9.97$ GHz, and $m = 3, 4, 5, 6$.

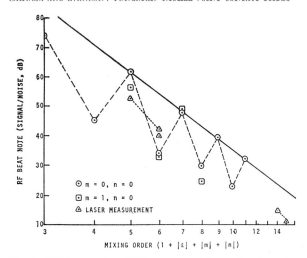

Fig. 1. RF beat note (S/N) versus mixing order for microwave signals at 20 GHz and laser signals at 3–11 THz. No bias voltage was used.

Laser frequency signals are indicated by triangles on Fig. 1. The triangles connected by the dashed line at mixing orders 5 and 6 are from $\nu_u = 32.1$ THz, $\nu_l = 10.7$ THz, $l = 3$, $n = 0$, and either $\nu_m = 20$ GHz, $m = 1$, or $\nu_m = 10$ GHz, $m = 2$. The lower triangle at the sixth order corresponds to $\nu_u = 3.8$ THz, $\nu_l = 0.964$ THz, $l = 4$, $\nu_m = 35$ GHz, $m = 1$, and $n = 0$. At the fourteenth and fifteenth mixing orders, the following laser signals are plotted: $\nu_u = 10.7$ THz, $\nu_l = 0.891$ THz, $l = 12$, $\nu_m = 29$ GHz, $m = 1$, and $n = 0$; and also $l = 12$, $m = 2$, $\nu_m = 14.564$, and $n = 0$.

At microwave frequencies the beat notes for the odd and even mixing orders form two distinct groups of data with the even mixing order signals lying 10–30 dB lower than the odd ones. It is the mixing order which denotes the even or odd character of the synthesized frequency as is shown by the three microwave signal case shown in Fig. 1. There are other differences between the two cases. In the even case, we sometimes observed two beat signal maxima as a function of power of the low frequency oscillator with the high frequency oscillator power kept at a low level. But when the high frequency oscillator power was increased, the low power maximum decreased and the high one increased. The latter is plotted in Fig. 1. In the odd case, the beat notes were more stable than in the even case where the peak value fluctuated several decibels. The largest signals from several different contacts are plotted in Fig. 1.

It has been reported that a dc bias increases the beat signal for the even order mixing [7]. A dc bias was applied after maximizing the beat note by adjusting the different microwave power levels. The dc bias increased the S/N of the even mixing case 15 dB at $n = 3$ and 7 dB at $n = 9$. In general, it decreased the signal for the odd mixing case. (We observed a slight increase under some conditions, but it was at most 1–2 dB.)

The S/N ratio for the odd microwave harmonics falls off roughly as the harmonic number to the minus 7.4 power. This rapid decrease precludes the synthesis of signals at high harmonics. A 0-dB signal (with a time constant of 1 ms) would be reached at the harmonic number 30, so that with some signal averaging, and sufficiently stable sources, a usable signal might be obtained at orders as high as 30–40.

At laser frequencies the odd-even difference was not observed; however, lack of sufficient data points makes it difficult to generalize. The laser data lie approximately between the envelope formed by the odd and even cases taken at microwave frequencies and there is very little difference between the odd and even cases. A dc bias increased the S/N for both the odd and even beat notes at the fifth and sixth orders, however, it was only by 2–4 dB (the bias had increased the microwave beat note by some 12 dB). A significant improvement at laser frequencies resulted from the use of a somewhat sharper tungsten wire and a lighter contact pressure. A somewhat higher dc impedance (100–200 Ω compared to 30–60 Ω with the coarser microwave contacts) and a larger rectified voltage were observed. It seems quite likely that the thinner–lighter contact has a significantly smaller capacitance, and consequently operates better at laser frequencies.

To test the diode's operation as a mixer at even higher frequencies, CW lasers, each oscillating on several modes at 1.06, 0.63, and 0.51 μm, radiated the diodes. In all cases the intermode beat signals were observed indicating that the diode was still operable as a mixer clear into the visible.

The responses of the microwave generated beat signals as a function of bias voltages are shown in Figs. 2–4. An even harmonic case (mixing order equal to 6) is shown in Figs. 2 and 3 for two different contacts with quite different impedances; the signal is very close to a minimum at zero bias. In Fig. 3 the zero bias is offset from center, and the oscillatory behavior of the diode is seen. The odd harmonic case with order 7 has its maximum approximately at zero bias, but otherwise possesses a similarly oscillatory behavior. The voltage spacing between minima of the beat signals depends strongly upon the contact pressure and the microwave power level. It becomes less as the microwave power is decreased.

At laser frequencies similar effects were seen; however, the minimum was no longer at zero bias in the even harmonic case, and there was less difference between the even and odd cases. Part of this effect might be due to self-biasing by the laser radiation.

To check the linearity of the beat signal versus the power of the lower basis frequency, the second harmonic of an X-band klystron was beat with a 19.98-GHz klystron with $\nu_{\text{beat}} \cong 20$ MHz. Fig. 5 shows the results. After maximizing the beat signal by varying the power of the higher frequency, we changed the power of the lower frequency with a calibrated X-band attenuator.

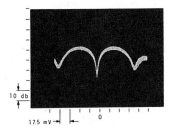

Fig. 2. RF beat note versus dc bias—even mixing case ($n = 5$).

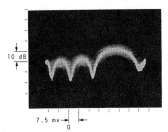

Fig. 3. RF beat note versus dc bias—even mixing case ($n = 5$) 0 bias displaced from center.

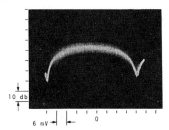

Fig. 4. RF beat note versus dc bias—odd mixing case ($n = 6$).

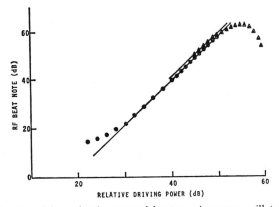

Fig. 5. RF beat signal generated by two microwave oscillators as a function of the lower frequency microwave oscillator power.

The power levels of the klystrons could be optimized independently. Sometimes, at high power, the diode condition changed and the beat intensity decreased many decibels. The erratic behavior of the diode necessitated the measurement first under low power and then under high power. The two sets of data in Fig. 5 show this result. The linear portion of this plot is represented by

$$\text{beat signal} \propto (\text{Drive power})^{+1.84}.$$

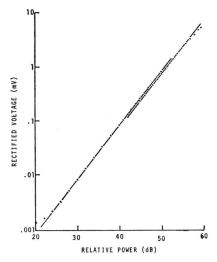

Fig. 6. Response of diode used as detector: rectified dc voltage versus microwave power.

The amount of drive power needed to generate the maximum RF beat note at different harmonics (with the high frequency power fixed) increased by 40 percent as the order was changed between 6 and 10.

The linearity of the rectified voltage from the detector was measured as a function of X-band power and is shown in Fig. 6. The detector is a linear device over three orders of magnitude (that is, the rectified voltage is proportional to the drive power). The voltage was detected with a calibrated phase sensitive detector and the X-band power was "chopped." The linearity of response indicates a small contribution from even terms of fourth order and higher in a power expansion of the voltage across the diode in terms of the applied radiation.

We also measured the dc response of the diode as a function of frequency and found a decrease of about an order of magnitude in going from 0.89 to 88 THz with about equal power from each laser. This decrease is a good deal smaller than the four orders of magnitude recently reported [16].

Conclusion

Several interesting characteristics of the metal–metal point contact diode have been revealed in this study, such as the oscillatory behavior shown in Fig. 3. It is hoped that some of these phenomena will be useful in finding an explanation for the physical processes occurring in the diode and also that they may be a guide to those simply wishing to use this device as a high-speed detector, harmonic generator, or mixer.

Acknowledgment

The authors wish to acknowledge the support of K. Gebert of the machine shop in fabricating the diode mounts, the able assistance of L. Mullen in building some of the various diodes tried, the discussions with E. Johnson, Jr., about the various theories of operation of point contact diodes, and the willing help of J. D. Cupp with the instrumentation.

One of the authors—E. Sakuma—especially appreciates the support and help given him by the staff of the Quantum Electronics Division while he was a guest worker for one year.

REFERENCES

[1] K. M. Evenson et al., "Accurate frequencies of molecular transitions used in laser stabilization: The 3.39-μm transition in CH_4 and the 9.33- and 10.18-μm transitions in CO_2," *Appl. Phys. Lett.*, vol. 22, p. 192, 1973.
[2] R. L. Barger and J. L. Hall, "Wavelength of the 3.39-μm laser-saturated absorption line of methane," *Appl. Phys. Lett.*, vol. 22, p. 196, 1973.
[3] K. M. Evenson et al., "Speed of light from direct frequency and wavelength measurements of the methane-stabilized laser," *Phys. Rev. Lett.*, vol. 29, p. 1346, 1972.
[4] J. W. Dees, "Detection and harmonic generation in the submillimeter wavelength region," *Microwave J.*, p. 48, Sept. 1966.
[5] V. Daneu, D. Sokoloff, A. Sanchez, and A. Javan, "Extension of laser harmonic-frequency mixing techniques into the 9 μ region with an infrared metal-metal point-contact diode," *Appl. Phys. Lett.*, vol. 15, p. 398, 1969.
[6] K. M. Evenson, G. W. Day, J. S. Wells, and L. O. Mullen, "Extension of absolute frequency measurements to the cw He-Ne laser at 88 THz (3.39 μ)," *Appl. Phys. Lett.*, vol. 20, p. 133, 1972.
[7] A. Sanchez, S. K. Singh, and A. Javan, "Generation of infrared radiation in a metal-to-metal point-contact diode at synthesized frequencies of incident fields: A high-speed broad-band light modulator," *Appl. Phys. Lett.*, vol. 21, p. 240, 1972.
[8] J. G. Simmons, "Generalized formula for the electric tunnel effect between similar electrodes separated by a thin insulating film," *J. Appl. Phys.*, vol. 34, p. 1793, 1963.
[9] S. P. Kwok, G. I. Haddad, and G. Lobov, "Metal-oxide-metal (M-O-M) detector," *J. Appl. Phys.*, vol. 42, p. 554, 1971.
[10] S. I. Green, "Point contact MOM tunneling detector analysis," *J. Appl. Phys.*, vol. 42, p. 1166, 1971.
[11] M. Nagae, "Response time of metal-insulator-metal tunnel junctions," *Japan. J. Appl. Phys.*, vol. 11, p. 1611, 1972.
[12] S. M. Faris, T. K. Gustafson, and J. C. Weisner, "Detection of optical and infrared radiation with DC-biased electron-tunneling metal-barrier-metal diodes," *IEEE J. Quantum Electron.*, vol. QE-9, pp. 737–745, July 1973.
[13] A. A. Lucas and P. H. Cutler, "Thermal field emission as a mechanism for infra-red laser light detection in metal whisker diode," presented at the 1st European Conf. Condensed Matter Summaries, Florence, Italy, Sept. 1971.
[14] E. G. Johnson, Jr., "DeBroglie wave scattering—A model for the metal-metal junction infrared frequency mixing process," private communication.
[15] J. R. Baird and R. Havemann, private communication.
[16] C. C. Bradley and G. J. Edwards, "Characteristics of metal-insulator-metal point-contact diodes used for two-laser mixing and direct frequency measurements," *IEEE J. Quantum Electron.* (Corresp.), vol. QE-9, pp. 548–549, May 1973.
[17] L. M. Matarrese and K. M. Evenson, "Improved coupling to infrared whisker diodes by use of antenna theory," *Appl. Phys. Lett.*, vol. 17, p. 8, 1970.

Speed of Light from Direct Frequency and Wavelength Measurements of the Methane-Stabilized Laser

K. M. Evenson, J. S. Wells, F. R. Petersen, B. L. Danielson, and G. W. Day
Quantum Electronics Division, National Bureau of Standards, Boulder, Colorado 80302

and

R. L. Barger* and J. L. Hall†
National Bureau of Standards, Boulder, Colorado 80302
(Received 11 September 1972)

> The frequency and wavelength of the methane-stabilized laser at 3.39 μm were directly measured against the respective primary standards. With infrared frequency synthesis techniques, we obtain $\nu = 88.376\,181\,627(50)$ THz. With frequency-controlled interferometry, we find $\lambda = 3.392\,231\,376(12)$ μm. Multiplication yields the speed of light $c = 299\,792\,456.2(1.1)$ m/sec, in agreement with and 100 times less uncertain than the previously accepted value. The main limitation is asymmetry in the krypton 6057-Å line defining the meter.

The speed of light is one of the most interesting and important of the fundamental (dimensioned) constants of nature.[1] It enters naturally into ranging experiments, such as geophysical distance measurements which use modulated electromagnetic radiation, and astronomical measurements such as microwave planetary radar and laser lunar ranging. Basically, very high-accuracy measured delay times for electromagnetic waves are dimensionally converted to distance using the light propagation speed. Recent experiments have set very restrictive limits on any possible speed dependence on direction[2] or frequency.[3] Another interesting class of applications involves the speed of propagating waves in a less obvious manner. For example, the conversion between electrostatic and electromagnetic units involves the constant c, as does the relativistic relationship between the atomic mass scale and particle energies.

With the perfection of highly reproducible and stable lasers, their wavelength-frequency duality becomes of wider interest. We begin to think of these lasers as *frequency* references for certain kinds of problems such as optical heterodyne spectroscopy.[4] At the same time, we use the wavelength aspect of the radiation, for example, in precision long-path interferometry.[5]

It has been clear since the early days of lasers that this wavelength-frequency duality could form the basis of a powerful method to measure the speed of light. However, the laser's optical frequency was much too high for conventional frequency measurement methods. This fact led to the invention of a variety of modulation or differential schemes, basically conceived to preserve the small interferometric errors associated with the short optical wavelength, while utilizing microwave frequencies which were still readily manipulated and measured. These microwave frequencies were to be modulated onto the laser output or realized as a difference frequency[6] between two separate laser transitions. Indeed, a proposed major long-path interferometric experiment[7] based on the latter idea has been made obsolete by the high-precision direct frequency measurement[8] summarized in this Letter. An ingenious modulation scheme, generally applicable to any laser transition, has recently successfully produced an improved value for the speed of light.[9] While this method can undoubtedly be perfected further, its differential nature leads to limitations which are not operative in the present, direct method.

The product of the frequency and wavelength of an electromagnetic wave is the speed of propagation of that wave. For an accurate determination of both of these quantities, the source should be stable and monochromatic and should be at as short a wavelength as possible. At shorter optical wavelengths the accuracy of the wavelength measurement increases. A suitable source of such radiation is the methane-stabilized He-Ne laser[10] at 3.39 μm (88 THz). Direct frequency measurements were recently extended to this frequency[11] and subsequently refined[8] to the present accuracy of 6 parts in 10^{10}. The wavelength of this stabilized laser has been compared[12] with the krypton-86 length standard to the limit of the usefulness of the length standard (approximately

3 parts in 10^9). The product of the measured frequency and the wavelength yields a new, definitive value for the speed of light, c. The previously accepted value[13] of c was similarly determined by measuring the frequency and wavelength of a stable electromagnetic oscillator; however, it oscillated at 72 GHz (more than 1000 times lower in frequency than in the case of the present measurements). The 100-fold improvement in the presently reported measurement comes mainly from the increased accuracy possible in the measurement[12] of the shorter wavelength.

With suitable point-contact mixer diodes, a chain of stabilized lasers and klystron frequency sources has allowed direct harmonic generation and frequency mixing from the National Bureau of Standards frequency standard upward to the CO_2 laser at 29 THz (10.3 μm) and thence upward to the methane-stabilized laser at 88 THz (3.39 μm). Five different types of laser and five klystrons were used in the three-step measurement process. An interpolating counter referenced to a cesium clock counted the X-band frequency at the base of the chain. The X-band klystron was phase locked to the 74-GHz klystron which was phase locked to the free-running HCN laser. A times-12 multiplication brought a harmonic of the HCN laser to within 29 GHz of the free-running H_2O laser frequency (at 28 μm), and the tens of megahertz beat note produced in this diode was measured on a spectrum analyzer and counter. The H_2O laser's third harmonic fell 19 GHz above the CO_2 $R(10)$ laser at 9.3 μm. The CO_2 laser was frequency stabilized to the central tuning dip (Lamb dip) of the saturated fluorescence in a low-pressure CO_2 absorption cell.[14,15] All of the above described beat notes were measured simultaneously in this first step of the experiment, yielding the frequency 32.134 266 891(24) THz for the $R(10)$ line. The interval from this $R(10)$ frequency to the $R(30)$ line of CO_2 at 10.3 μm was measured as the HCN laser's third harmonic +19.5 GHz. The resulting CO_2 $R(30)$ frequency was 29.442 483 315(25) THz. The third harmonic of this laser's frequency falls 49 GHz short of the He-Ne laser (3.39 μm) stabilized to the saturated absorption peak in methane. The final methane-stabilized frequency [$F_1^{(2)}$ component of $P(7)$] was found to be 88.376 181 627(50) THz. The fractional uncertainties in the molecular frequencies of CO_2 are somewhat larger than that of methane due to larger possible offsets from true line centers of the CO_2 absorptions.

These offsets would not affect either the value or the uncertainty of the measured methane frequency. The (1 standard deviation) error estimates result from careful analysis[8] of both random and possible systematic effects. The dramatic accuracy improvement over previous infrared frequency measurements stems from the use of better microwave and laser frequency control and measurement electronics, improved mixer signal-to-noise ratios, and, most importantly, the use of molecular saturated-absorption stabilization of the measured infrared frequencies.

In a coordinated effort, the wavelength of the 3.39-μm line of methane has been measured with respect to the Kr^{86} 6057-Å primary standard of length. Using a frequency-controlled Fabry-Perot interferometer with a pointing precision of about 2×10^{-5} orders, we have made a detailed search for systematic offsets inherent in the experiment, including effects due to the asymmetry of the Kr standard line. Offsets due to various experimental effects (such as beam misalignments, mirror curvatures and phase shifts, phase shift over the exit aperture, diffraction, etc.) were carefully measured and then removed from the data with an uncertainty of about 2 parts in 10^9. This reproducibility for a single wavelength measurement illustrates the high precision which is available using the frequency-controlled interferometer.

Unfortunately, after the Kr^{86} transition at 6057 Å was adopted as the primary standard of length it was discovered that this line is slightly asymmetric,[16] resulting in a small shift of effective wavelength with the order of interference. For example, in our experiment[12] the apparent measured wavelength showed a fractional systematic dependence of $\pm 1.1 \times 10^{-8}$ upon the mirror spacing, in basic agreement with other work.[16,17] Following Rowley and Hamon,[16] a two-component model of the krypton asymmetry was used to analyze our data. This model reduced the standard deviation of the twenty wavelength measurements from 6.4 to 2.7 parts in 10^9, and shifted the point of fringe maximum intensity by 4.1 parts in 10^9. The deviations were improved somewhat further and the average wavelength red-shifted by 1.2 parts in 10^9 when we also assumed a radial dependence[18] of the Doppler shift[19] across the capillary bore of the krypton standard lamp.

In view of the (small) intrinsic asymmetry of the Kr standard line, it is necessary to specify the point on the line profile to which the defined wavelength (6057.802 105 Å) is applied. At present there is no universal convention for this choice. Thus if the defined value is applied to the maximum-intensity point of the Kr line, we find λ = 33 922.314 04 Å; if the defined value is applied to the center of gravity of the Kr line, λ = 33 922.313 76 Å. Detailed consideration[20] of random and known systematic effects, along with uncertainties in the krypton asymmetry model, leads to an estimated 68% confidence interval of $\delta\lambda = \pm 1.2 \times 10^{-4}$ Å or $\delta\lambda/\lambda = \pm 3.5 \times 10^{-9}$ for both of these results.

The methane wavelength has also been measured by Giacomo.[21] Although he does not state his reference-point convention, his quoted result (33 922.313 76 Å) is identical to ours for our case where the defined Kr wavelength is applied to the line center of gravity.

In the absence of an international agreement on this question of reference point for the krypton length definition, we feel it simplifies presentation of our numerical result if we adopt the arbitrary convention that the defined wavelength is to be applied to the center of gravity of the krypton line. With this choice, the values of the wavelength and frequency of the methane-stabilized He-Ne laser are

$$\lambda = 3.392\,231\,376(12)\ \mu m \quad (\delta\lambda/\lambda = \pm 3.5 \times 10^{-9})$$

and

$$\nu = 88.376\,181\,627(50)\ \text{THz} \quad (\delta\nu/\nu = \pm 6 \times 10^{-10}).$$

Therefore,

$$c = 299\,792\,456.2(1.1)\ \text{m/sec}$$
$$(\delta c/c = \pm 3.5 \times 10^{-9}).$$

The uncertainties quoted are 1-standard-deviation (68% reliance) estimates and include both random and residual systematic uncertainties. This result is in agreement with the previously accepted value of $c = 299\,792\,500(100)$ m/sec and is about 100 times more accurate. As mentioned above, a recent differential measurement of the speed of light has been made by Bay, Luther, and White[9]; their value is 299 792 462(18) m/sec, which is in agreement with the presently determined value. If the maximum-intensity point of the Kr line is chosen,[22] the methane wavelength, and hence the value of c, is increased by 8.3 parts in 10^9 [$c_{\text{max}\,I} = 299\,792\,458.7(1.1)$ m/sec].

The fractional uncertainty in our value for the speed of light, $\pm 3.5 \times 10^{-9}$, essentially arises from the interferometric measurements with the incoherent krypton radiation which operationally defines the international meter. This limitation is indicative of the remarkable growth in optical physics in recent years: The present krypton-based length definition was adopted only in 1960!

One view of this situation is that with lasers (and *great* care) it is possible to measure optical lengths more precisely than they may be operationally expressed in meters. Thus, one is easily led to consider choosing a suitable stabilized laser as a new basic standard of length. Both the methane-stabilized[10,23] He-Ne laser at 3.39 μm (88 THz) and the I_2-stabilized[24] He-Ne laser at 0.633 μm appear to be suitable candidates for the basic standard of length. They also can serve as secondary standards of frequency in the near-infrared and visible regions. The methane-stabilized He-Ne laser frequency is already known to 6 parts in 10^{10}, and further measurements are expected to increase the accuracy to a few parts in 10^{11} in the next year or two. A new value of the speed of light with this accuracy should thus be achievable if the standard of length were redefined.

Alternately, one can consider defining the meter as a specified fraction of the distance light travels in one second in vacuum (that is, one can define the speed of light). With this definition, the wavelength of stabilized lasers would be known to the same accuracy with which their frequencies can be measured. Stabilized lasers would thus provide accurate secondary standards of both frequency and length. It should be noted that an adopted nominal value for the speed of light is already in use for high-accuracy astronomical measurements.

Independent of which type of definition is chosen we believe that research on simplified frequency synthesis chains bridging the microwave-optical gap will be of great interest, as will refined experiments directed toward an understanding of the factors that limit laser optical frequency reproducibility. No matter how such research may turn out, it is clear that ultraprecise physical measurements made in the interim can be preserved through wavelength or frequency comparison with a suitable stabilized laser such as the 3.39-μm methane device.

The authors would especially like to thank P. L. Bender for his long-term interest in laser speed-of-light experiments, for his encouragement and enthusiasm, and for discussions of many aspects of this measurement. We also express our gratitude to those whose work with the

Josephson junction was a parallel effort to ours in trying to achieve near-infrared frequency measurements. D. G. McDonald's leadership in that effort has been extremely valuable. A. S. Risley's work on improving the spectral purity of the X-band sources for the Josephson junction attempts have been very useful to us. We are grateful to J. D. Cupp for his advice and his excellent electronic support. We note here that experiments using the Josephson junction to measure laser frequencies are continuing, and may well lead to better methods for near-infrared frequency synthesis in the future. The collaboration with H. S. Boyne in earlier phases of this work and his continued interest and support of this effort are appreciated. Also we extend special appreciation to P. Giacomo of the Bureau International des Poids et Mesures for useful discussions with two of us (R.L.B. and J.L.H.) and for his very helpful criticisms and comments regarding the wavelength determination.

---

*Quantum Electronics Division.

†Laboratory Astrophysics Division, and Joint Institute for Laboratory Astrophysics (operated jointly by the National Bureau of Standards and University of Colorado).

[1]The interested reader will find a useful, critical discussion of the speed of light in D. D. Froome and L. Essen, *The Velocity of Light and Radio Waves* (Academic, New York, 1969).

[2]T. S. Jaseja, A. Javan, J. Murray, and C. H. Townes, Phys. Rev. 133, A 1221 (1964), using infrared masers; D. C. Champeney, G. R. Isaak, and A. M. Khan, Phys. Lett. 7, 241 (1963), using Mössbauer effect.

[3]B. Warner and R. E. Nather, Nature (London) 222, 157 (1969), from dispersion in the light flash from pulsar NP 0532, obtain $\Delta c/c \leq 5 \times 10^{-18}$ over the range $\lambda = 0.25$ to 0.55 μm.

[4]E. E. Uzgiris, J. L. Hall, and R. L. Barger, Phys. Rev. Lett. 26, 289 (1971).

[5]J. Levine and J. L. Hall, J. Geophys. Res. 77, 2595 (1972).

[6]J. L. Hall and W. W. Morey, Appl. Phys. Lett. 10, 152 (1967).

[7]J. Hall, R. L. Barger, P. L. Bender, H. S. Boyne, J. E. Faller, and J. Ward, Electron Technol. 2, 53 (1969).

[8]K. M. Evenson, J. S. Wells, F. R. Petersen, B. L. Danielson, and G. W. Day, to be published.

[9]Z. Bay, G. G. Luther, and J. A. White, Phys. Rev. Lett. 29, 189 (1972).

[10]R. L. Barger and J. L. Hall, Phys. Rev. Lett. 22, 4 (1969).

[11]K. M. Evenson, G. W. Day, J. S. Wells, and L. O. Mullen, Appl. Phys. Lett. 20, 133 (1972).

[12]R. L. Barger and J. L. Hall, to be published.

[13]K. D. Froome, Proc. Roy. Soc., Ser. A 247, 109 (1958).

[14]C. Freed and A. Javan, Appl. Phys. Lett. 17, 53 (1970).

[15]F. R. Petersen and B. L. Danielson, to be published.

[16]W. R. C. Rowley and J. Hamon, Rev. Opt., Theor. Instrum. 42, 519 (1963).

[17]This ^{86}Kr reproducibility limit is just larger than the 1×10^{-8} stated in *Comité Consultatif pour la Definition du Metre, Rapport, 1970* (Bureau International des Poids et Mesures, Sèvres, France, 1972).

[18]A radial variation of the Doppler shift was postulated in 1963 by F. Bayer-Helms of the Physikalische-Technische Bundesanstalt (private communication to R.L.B. and J.L.H.).

[19]K. M. Baird and D. S. Smith, Can. J. Phys. 37, 832 (1957). See also Ref. 14.

[20]For a more detailed discussion of asymmetry corrections, see R. L. Barger and J. L. Hall, to be published.

[21]P. Giacomo, in *Proceedings of Fourth International Conference on Atomic Masses and Fundamental Constants, Teddington, England, September 1971* (Plenum, New York, 1972).

[22]In the absence of an international agreement on this question, we slightly prefer the center-of-gravity definition since it probably would be less affected by lamp operating conditions (influence on Doppler width).

[23]J. L. Hall and R. L. Barger, in Proceedings of the Symposium on Basic and Applied Laser Physics, Esfahan, 1971 (Wiley, New York, to be published).

[24]G. R. Haines and C. E. Dahlstrom, Appl. Phys. Lett. 14, 362 (1969); G. R. Haines and K. M. Baird, Metrologia 5, 32 (1969).

Accurate frequencies of molecular transitions used in laser stabilization: the 3.39-μm transition in CH_4 and the 9.33- and 10.18-μm transitions in CO_2

K.M. Evenson, J.S. Wells, F.R. Petersen, B.L. Danielson, and G.W. Day

Quantum Electronics Division, National Bureau of Standards, Boulder, Colorado 80302
(Received 10 November 1972)

The frequencies of three lasers stabilized to molecular absorptions were measured with an infrared-frequency synthesis chain extending upwards from the cesium frequency standard. The measured values are 29.442 483 315(25) THz for the 10.18-μm $R(30)$ transition in CO_2, 32.134 266 891(24) THz for the 9.33-μm $R(10)$ transition in CO_2, and 88.376 181 627(50) THz for the 3.39-μm $P(7)$ transition in CH_4. The frequency of methane, when multiplied by the measured wavelength reported in the following letter, yields 299 792 456.2(1.1) m/sec for the speed of light.

The direct measurement of frequencies was recently extended to the 88-THz (3.39-μm) He-Ne laser.[1] Prior to this measurement, frequencies of the HCN,[2] H_2O,[3] and CO_2[4] lasers had been measured, completing a chain of frequency measurements extending from the frequency standard. In these previous measurements the lasers were tuned to the peaks of their gain curves; consequently, accuracy was limited to a few parts in 10^7. This letter describes remeasurements of differences between these lasers using CO_2 and He-Ne lasers stabilized with saturated molecular absorption in CO_2[5] and CH_4,[6] and presents results with an increase in accuracy of more than a factor of 100. The wavelength of the same $P(7)$ methane transition used to stabilize the He-Ne laser has been measured,[7] and the product of the new value for frequency and wavelength yields an updated value for the speed of light considerably more accurate than the presently accepted value. The new value of c is limited only by the present length standard.[7]

A block diagram in Fig. 1 illustrates the entire laser frequency chain. The three saturated-absorption-stabilized lasers are shown in the upper right-hand section, and the transfer chain oscillators are in the center column. The He-Ne and CO_2 lasers in the transfer chain were offset locked[8]; that is, they were locked at a frequency a few megahertz different from the stabilized lasers. This offset-locking procedure produced He-Ne and CO_2 transfer oscillators without the frequency modulation used in the molecular-stabilized lasers. The measurements of the frequencies in the entire chain were made in three steps shown on the right-hand side, by using standard heterodyne techniques previously described.[1–4]

Conventional silicon point-contact harmonic generator-mixers were used up to the frequency of the HCN laser. Above this frequency, tungsten-on-nickel diodes were used as harmonic generator-mixers. These metal-metal diodes required 50 or more mW of power from the lasers. The 2-mm-long 25-μm-diam tungsten antenna, with a sharpened tip which lightly contacted the nickel surface, seemed to couple to the radiation in two separate manners. At 0.89 and 10.7 THz it acted like a long wire antenna,[9,10] while at 29–88 THz its conical tip behaved like one-half of a biconical antenna.[10] Conventional detectors were used in the offset-locking steps.

The methane-stabilized He-Ne laser used in these experiments is quite similar in size and construction to the device described by Hall and Barger.[6] The gain tube was dc excited, and slightly higher reflectivity mirrors were employed. The latter resulted in a higher energy density inside the resonator and consequently a somewhat broader saturated absorption. Pressure in the internal methane absorption cell was about 0.01 Torr (1 Torr = 133.3 N/m^2).

The two 1.2-m-long CO_2 lasers used in the experiments contained internal absorption cells and dc-excited sealed gain tubes. A grating was used on one end for line selection, and frequency modulation was achieved by dithering the 4-m–radius-of-curvature mirror on the oppo-

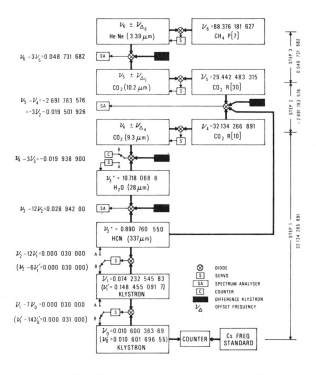

FIG. 1. Stabilized laser frequency synthesis chain. All frequencies are given in THz; those marked with an asterisk were measured with a transfer laser oscillator tuned to approximate line center.

TABLE I. Summary of Measurements. Molecular frequencies (the CO_2 frequencies include a 20-kHz systematic uncertainty combined with previous uncertainties in ν_4 and ν_5): $\nu_4 = 32.134\,266\,891(24)$, $\nu_5 = 29.442\,483\,315(25)$, $\nu_6 = 88.376\,181\,627(50)$.

Run	Step 1 ν_4	Step 2 $\nu_5 - \nu_4$ [a]	Step 3 $\nu_6 - 3\nu_5$ [b]
I	32.134 266 885 (37)	−2.691 783 577 (9) −2.691 783 573 (9)	0.048 731 658 (65)
II	32.134 266 925 (31)	−2.691 783 595 (14)	0.048 731 656 (65)
III	32.134 266 885 (20)	−2.691 783 577 (9) −2.691 783 573 (9)	0.048 731 693 (25)
IV	32.134 266 890 (14)	−2.691 783 574 (9) −2.691 783 575 (9)	0.048 731 680 (20)
Weighted values	32.134 266 891 (14),	−2.691 783 576 (5)	0.048 731 682 (20)

[a]Double entries correspond to interchanging the two CO_2 lasers. [b]Includes a −12-kHz correction to runs III and IV.

site end. CO_2 pressure in the internal absorption cell was 0.020 Torr. The laser frequency was locked to the zero-slope point on the Lamb dip in the 4.3-μm fluorescent radiation.[5] The 0.89-, 10.7-, and 88-THz transfer lasers were 8-m-long linearly polarized cw oscillators with single-mode output power greater than 50 mW. The Michelson HCN laser has been described.[11] The H_2O laser used a double-silicon-disk partially transmitting end mirror and a 0.5-mil polyethylene internal Brewster-angle membrane served to polarize the laser. The 8-m He-Ne laser oscillated in a single mode without any mode selectors because of a 4-Torr pressure with a 7:1 ratio of helium to neon. This resulted in a pressure width approximately equal to the Doppler width, and the high degree of saturation allowed only one mode to oscillate.

Conventional klystrons used to generate the four difference frequencies between the lasers were all stabilized by standard phase-lock techniques, and their frequencies were determined by cycle counting at X band.

An interpolating counter controlled by a cesium clock via the AT (NBS) time scale[12,13] in the NBS Time and Frequency Division counted the 10.6-GHz klystron in the transfer chain. This same standard was used to calibrate the other counters and the spectrum analyzer-tracking generator.

In step 1, a frequency synthesis chain was completed from the cesium standard to the stabilized $R(10)$ CO_2 laser. All difference frequencies in this chain were either measured simultaneously or held constant. Each main chain oscillator had its radiation divided so that all beat notes in the chain could be measured simultaneously. For example, a silicon-disk beam splitter divided the 10.7-THz beam into two parts: one part was focused on the diode which generated the 12th harmonic of the HCN laser frequency, the remaining part irradiated another diode which generated its own third harmonic and mixed with the output from the 9.3-μm CO_2 laser and the 20-GHz klystron.

Figure 1 shows the two different ways in which the experiment was carried out. In the first scheme (output from mixers in position A), the HCN laser was frequency locked to a quartz crystal oscillator via the 148- and 10.6-GHz klystrons, and the frequency of the 10.6-GHz klystron was counted. The H_2O laser was frequency locked to the stabilized CO_2 laser, and the beat frequency between the H_2O and HCN lasers was measured on the spectrum analyzer. In the second scheme (output from mixers in position B) the 10.6-GHz klystron was phase locked to the 74-GHz klystron, which in turn was phase locked to the free-running HCN laser. The 10.6-GHz klystron frequency was again counted. The free-running H_2O laser frequency was monitored relative to the stabilized CO_2 laser frequency. The beat frequency between the H_2O and HCN lasers was measured as before on the spectrum analyzer.

In step 2, the difference between the two CO_2 lines was measured. The HCN laser remained focused on the diode used in step 1, which now also had two CO_2 laser beams focused on it. The sum of the third harmonic of the HCN frequency, plus a microwave frequency, plus the measured rf beat signal is the difference frequency between these two CO_2 lines. The two molecular-absorption-stabilized CO_2 lasers were used directly, and the relative phase and amplitudes of the modulating voltages were adjusted to minimize the width of the beat note. The beat note was again measured on a combination spectrum analyzer and tracking generator-counter. Dual entries in step 2 of Table I arose from interchanging the roles of the CO_2 lasers to detect possible systematic differences in the two laser-stabilization systems.

In step 3, the frequency of the $P[7]$ line in methane was measured relative to the $R(30)$ line (10.4-μm band) of CO_2. Both the 8-m 3.39-μm laser and the CO_2 laser were offset locked from saturated-absorption-stabilized lasers and thereby not modulated. The 10- to 100-MHz beat note was again measured either on a spectrum analyzer and tracking generator, or, in the final measurement when the S/N ratio of the beat note was large enough (about 100), directly on a counter.

The measurements are chronologically divided into four runs, and the results are presented in Table I along with a one-standard-deviation estimation of the uncertainties. Efforts were made to make experimental improvements from one run to the next, and consequently the results from run IV have the smallest experimental uncertainties. Run II contains results of measurements on two days, in which all three steps were accomplished each day. The results were marred, however, by a leaky CO_2 absorption cell, which may account for the differences with respect to runs I, III, and IV. In runs I, III, and IV, the various steps were sometimes per-

TABLE II. Estimated uncertainties in the data of run IV. Parts in 10^{10}.

Term	Measurement	Uncertainty in transfer oscillator frequencies
Step 1		
ν_0	Cs standard	0.7
ν_0	±1 count	0.001
$\nu_0 \to \nu_1$ lock frequency	SA-TG[a]	1.2
$\nu_1 \to \nu_2$ lock frequency	SA-TG	0.1
$\nu_2 \to \nu_3$ beat note	SA-TG	2.0
$\nu_2 \to \nu_3$ beat note	Possible asymmetry[b]	3.0
$\nu_3 \to \nu_4$ beat note	Counter (±1 count)	0.3
ν_4 resetability		2.0
$\nu_0 \to \nu_4$ statistical fluctuations	Standard deviation mean ($N=25$)	0.7
Step 1: σ_{ν_4}	$(\Sigma_i \sigma_i^2)^{1/2}$ (total of above)	4.4
Step 2		
$\nu_4 \to \nu_5$ beat note	Possible asymmetry[b]	1.0
$3\nu_2$	First four terms in step 1 divided by 10	0.14
ν_4 resetability		2.0
ν_5 resetability		2.0
$\nu_4 \to \nu_5$ statistical fluctuations	Standard deviation mean ($N=20$)	0.3
Step 2: $\sigma_{\nu_4 \to \nu_5}$	$(\Sigma_i \sigma_i^2)^{1/2}$	3.0
Step 3		
$\nu_5 \to \nu_6$ beat note	Counter	0.1
ν_5 resetability		2.0
ν_6 resetability		1.0
$\nu_5 \to \nu_6$ statistical fluctuations	Standard deviation mean ($N=20$)	0.2
Step 3: $\sigma_{\nu_5 \to \nu_6}$	$(\Sigma_i \sigma_i^2)^{1/2}$	2.2

[a]SA-TG, spectrum analyzer-tracking generator.   [b]4% of linewidth.

formed on different days, but all were done within several days of each other. The cumulative changes in experimental techniques led to the improved results in run IV. With configuration B (step 1 in Fig. 1) an improved signal-to-noise ratio facilitated connection of the 74-GHz klystron and the HCN laser with a phase-lock loop and thus decreased the uncertainty in the HCN laser frequency. The accuracy of the H_2O to CO_2 $R(10)$ beat frequency measurement was improved by cycle counting, as was the CO_2 $R(30)$ to He-Ne beat frequency measurement in step 3.

Weighted values for each of the steps and for ν_4, ν_5, and ν_6 were obtained by weighting the results of all runs inversely as the square of the standard deviations.

Table II lists the individual uncertainties in various parts of the frequency synthesis chain as it was used during the final set of measurements (run IV). One notes that the largest uncertainty is a possible asymmetry in the ν_2-to-ν_3 beat note. The error listed for this factor is two-thirds of the estimated limit of detectability of such an asymmetry. This asymmetry might arise, for example, from a nonsymmetrical excursion of the H_2O and HCN laser mirrors.

The uncertainties in the weighted values for steps 1 and 3 were taken to be the same as those of run IV; this was done because of the uncertainties, such as the possible asymmetry in the beat note between ν_2 and ν_3, might be of the same sign from run to run. In step 2, however, the major uncertainties were independent, and the final uncertainty was obtained in the normal fashion.

Numerous other possible errors were considered; however, they were all negligible compared with those listed. Some of those considered were drift of the HCN and H_2O lasers during the measurement, frequencies of difference klystrons, and asymmetries in other beat notes (these were eliminated by direct counting). No frequency shifts are known to result from the multiplication process in the metal-metal diodes.

The saturated-methane line shape and resetability (which can be associated with various kinds of frequency shifts, such as pressure, power, and base line slope) of the stabilized He-Ne laser have been studied in some detail by Hall and Barger.[6] Since the characteristics of our device were not known as well, a comparison was made with a Hall-Barger laser via a transfer laser. A 12-kHz difference, which remained after a base line slope correction had been made, probably resulted from an asymmetric line in our laser caused by too much power inside the resonator. This difference was applied as a correction to the results and appears in Table I. The 1×10^{-10} resetability error assigned to ν_6 (Table II) occurs largely because of the uncertainties involved in transferring the accuracy of the Hall-Barger laser to our laboratory. No corrections were applied for pressure or shifts other than those mentioned, which are believed to be small compared to the assigned resetability error.

Corresponding information for the stabilized CO_2 laser is currently being investigated in our laboratory and will be reported in a future publication.[14] Some, if not all, of the factors which affect the symmetry and frequency of the methane line also affect the saturated CO_2 lines. No correction for sloping base line, pressure, power, or other shifts was made in the CO_2 frequencies in Table I. From preliminary measurements, it is estimated that base line and power shifts are less than 20 kHz. The pressure shift is believed to be negative and less than 100 Hz/mTorr.

The accuracy of the methane measurement depends on the stability and resetability of the CO_2 lasers rather than on the absolute accuracy of their frequencies since the CO_2 lasers served as transfer oscillators. The fractional frequency variation is $3 \times 10^{-11} \tau^{-1/2}$ for $10^{-2} \leq \tau \leq 10$ sec. It is estimated that the resetability of each CO_2 laser is about 2×10^{-10} (Table II). The stability characteristic is reflected in the statistical fluctuations in Table II.

Since the shifts affecting the true molecular CO_2 frequencies have not been measured, an additional uncertainty of about 20 kHz has been included for these frequencies, as shown in Table I.

Frequencies are currently measurable to parts in 10^{13}, and hence the over-all error of about six parts in 10^{10} represents a result which can be improved upon. The experiment was done fairly quickly to obtain frequency measurements of better accuracy than the wavelength measurements; this was easily done. It should be possible to obtain 50—100 times more accuracy by using tighter locks on the lasers (such as phase locking the HCN laser[15,16]) and by using counting techniques at all beat notes.

The relative ease with which these laser harmonic signals were obtained in this second round of frequency measurements indicates that the measurement of the frequencies of visible radiation now appears very near at hand. Such measurements should greatly facilitate one's ability to accurately utilize the visible and infrared portion of the electromagnetic spectrum.

A measurement of the wavelength of methane reported in the following letter[7] yields a value of 33 922.313 76 (12) Å ($\delta\lambda/\lambda = \pm 3.5 \times 10^{-9}$) when referenced to the center of gravity of the krypton length standard. This value multiplied by the measured frequency, 88.376 181 627 (50) THz ($\delta f/f = \pm 5.6 \times 10^{-10}$), yields a definitive value for the speed of light of $c = 299\ 792\ 456.2\ (1.1)$ m/sec ($\delta c/c = 3.5 \times 10^{-9}$). This number is in agreement with the previously accepted value[17] of 299 792 500 (100) m/sec and is about 100 times more accurate. A recent differential measurement of the speed of light has been made by Bay, Luther, and White[18]; their value is 299 792 462 (18) m/sec, which also is in agreement with the presently determined value.

The uncertainty in our value for the speed of light essentially arises from the interferometric measurements with the incoherent krypton radiation which operationally defines the international meter. Some various alternatives to the present length standard are discussed elsewhere.[19]

The authors are indebted to numerous individuals for contributions to the work reported here. D. A. Jennings wholeheartedly supported the work from its inception and provided valuable suggestions. The enthusiasm of A.L. Schmeltekopf and many discussions with him have played a significant role in these measurements. L.B. Elwell has been indispensable to the project. He and his support group, J.J. Skudler and K. Rosner, have contributed uniquely designed hardware and service which permitted this experiment to progress unimpeded. The assistance of J. Hall and H. Hellwig on problems associated with the methane-stabilized laser is appreciated, as is the consultation with J. Shirley on the physics of saturated-absorption processes. The authors would like to thank D.G. McDonald, whose parallel efforts with the Josephson junction have contributed to the progress of the experiment. The authors have had numerous stimulating discussions on time and frequency standards with D. Halford and are grateful not only to him but also to A. Risley and J. Shoaf for contributions and cooperation. In addition to his long-term interest in the project, P. Bender also provided valuable assistance in the error analysis. Eiichi Sakuma's assistance many times during the experiment is sincerely appreciated. Vernon Derr and R. Strauch have provided stimulating discussions and assistance in some of the earlier frequency measurements culminating this experiment. Thanks are also due to L.O. Mullen for assistance on diode research, J.D. Cupp for technical support, and to three excellent instrument makers, J. Wichman, K. Gebert, and W. Jackson.

---

[1]K.M. Evenson, G.W. Day, J.S. Wells, and L.O. Mullen, Appl. Phys. Lett. 20, 133 (1972).
[2]L.O. Hocker, A. Javan, D. Ramachandra Rao, L. Frenkel, and T. Sullivan, Appl. Phys. Lett. 10, 5 (1967).
[3]K.M. Evenson, J.S. Wells, L.M. Matarrese, and L.B. Elwell, Appl. Phys. Lett. 16, 159 (1970).
[4]K.M. Evenson, J.S. Wells, and L.M. Matarrese, Appl. Phys. Lett. 16, 251 (1970).
[5]Charles Freed and Ali Javan, Appl. Phys. Lett. 17, 53 (1970).
[6]J. L. Hall in *Esfahan Symposium on Fundamental and Applied Laser Physics*, edited by M. Feld and A. Javan (Wiley, New York, to be published).
[7]R.L. Barger and J.L. Hall, following letter, Appl. Phys. Lett. 22, 196 (1973).
[8]R.L. Barger and J.L. Hall, Phys. Rev. Lett. 22, 4 (1969).
[9]L.M. Matarrese and K.M. Evenson, Appl. Phys. Lett. 17, 8 (1970).
[10]*Antenna Engineering Handbook*, edited by Henry Jasik (McGraw-Hill, New York, 1961), Chap. 4 and 10.
[11]K.M. Evenson, J.S. Wells, L.M. Matarrese, and D.A. Jennings, J. Appl. Phys. 42, 1233 (1971).
[12]D.W. Allan, J.E. Gray, and H.E. Machlan, IEEE Trans. Instrum. Meas. **IM-21**, 388 (1972).
[13]Helmut Hellwig, Robert F.C. Vessot, Martin W. Levine, Paul W. Zitzewitz, David W. Allan, and David Glaze, IEEE Trans. Instrum. Meas. **IM-19**, 200 (1970).
[14]F.R. Petersen and B.L. Danielson (unpublished).
[15]R.E. Cupp, V.J. Corcoran, and J.J. Gallagher, IEEE J. Quantum Electron. **QE-6**, 241 (1970).
[16]J.S. Wells, IEEE Trans. Instrum. Meas. (to be published).
[17]K.D. Froome, Proc. R. Soc. Lond. 247A, 109 (1958).
[18]Z. Bay, G.G. Luther, and J.A. White, Phys. Rev. Lett. 29, 189 (1972).
[19]K.M. Evenson, J.S. Wells, F.R. Petersen, B.L. Danielson, G.W. Day, R.L. Barger, and J.L. Hall, Phys. Rev. Lett. 29, 1346 (1972).

Reprinted with permission from *Applied Physics Letters* 22, 196–199, ©1973 The American Physical Society.

Wavelength of the 3.39-μm laser-saturated absorption line of methane

R.L. Barger
Quantum Electronics Division, National Bureau of Standards, Boulder, Colorado 80302

J.L. Hall
*Laboratory Astrophysics Division, National Bureau of Standards, Boulder, Colorado 80302
and Joint Institute for Laboratory Astrophysics,* Boulder, Colorado 80302*

(Received 10 November 1972)

The wavelength of the 3.39-μm line of methane has been measured with respect to the Kr^{86} 6057-Å standard by using a frequency-controlled Fabry-Perot interferometer. We have exhaustively studied systematic offsets inherent in the experiment, including effects due to asymmetry of the Kr standard line. Lacking a convention relating the defined Kr wavelength 6057.802 105 Å to observables of the krypton line (e.g., center of gravity or fringe maximum intensity point), we report two methane wavelengths: $\lambda_{max\,I}$ = 33 922.314 04 Å and λ_{cg} = 33 922.313 76 Å. Both results have an uncertainty of $\delta\lambda$ = ±1.2×10^{-4} Å or $\delta\lambda/\lambda$ = ±3.5 ×10^{-9}. Multiplication by the frequency measurement of the preceding letter gives the speed of light, c_{cg} = 299 792 456.2(1.1) m/sec.

The laser-saturated absorption line of methane at 3.39 μm [the $P(7)$ line of the ν_3 band] has been shown to have remarkable characteristics desired of radiation which might be used as a standard of length (or frequency).[1] The Zeeman coefficient is small[2] (+0.31 μN), and the transition is free from first-order Stark shift.[3] Linewidths as narrow as 25 kHz have been achieved. This, together with the signal-to-noise ratio of several thousand attainable with laser radiation, results in an achieved[4] frequency stability of a few parts in 10^{14}. Line center has been shown[1] reproducible to better than 1×10^{-11}. The selective-collision process in saturated absorption results in a pressure shift to broadening ratio of less than about 1:200 rather than the approximately 1:4 produced in van der Waals interactions. With a half-width half-maximum (HWHM) of 25 kHz, the pressure broadening (of methane by methane) is 10 kHz/mTorr, and the shift is $\leq$ 50 Hz/mTorr.[4]

These properties of the molecule and the saturation process, together with the laser's high intensity and spatial coherence, result in the 3.39-μm radiation having the most accurately measurable wavelength obtained to date. Thus it is of interest to measure the methane wavelength as accurately as possible in terms of the international standard of length, the meter, which is defined by the Kr^{86} orange line at 6057.802 105 Å. Long before a consensus emerges on the optimum technical route for redefinition of the meter, we believe the precision measurement community will find the methane transition an attractive and useful wavelength reference. One example of the new precision measurement possibilities is the extension of the range of direct frequency measurements to the 3.39-μm line.[5] This frequency value, taken with the presently reported wavelength determination, results in a definitive value for the speed of light as reported here and discussed elsewhere.[6]

The methane-stabilized laser used in this measurement had an intracavity methane absorption cell. The 10-mTorr pressure of methane in the cell together with the laser mode radius of 0.75 mm resulted in a saturation peak with an intensity of 3% of the laser power and a half-width half-maximum of 300 kHz (1×10^{-5} cm^{-1}, line Q of 1.4×10^8). Thus, the desired measurement accuracy of a few parts in 10^9 for $\Delta\nu/\nu$, which is the practical limit for the krypton standard, corresponds to a measurement of methane line center to only within the line half-width.

The Kr^{86} standard lamp was the type described by Englehard and Terrien.[7] The tube contained Kr^{86} with an isotopic purity of 99.8% quoted by the manufacturer. The discharge was run with a current density of 0.3 A/cm^2 with the tube immersed in a bath of liquid nitrogen held at the triple point. The observed light travels from cathode toward anode. The 6057-Å line emitted from a Kr lamp run under these conditions has a defined accuracy of one part in 10^8, but it is possible to measure a point on the line, and to hold the systematic errors associated with this point, to a few parts in 10^9 as has been done in our measurements. As a precaution against an unknown systematic error being associated with our lamp, we have compared the 6057-Å line from our lamp with that from a lamp loaned to us by another laboratory.[8] The two wavelengths were offset by 1 ± 2.5 parts in 10^9, well within the accuracy limit for these lamps. To make such precise measurements, it is essential to deal effectively with the small asymmetry of the krypton emission-line profile.

We have developed and used for the measurement reported here a new interferometry technique[9] wherein the interferometer length is servo controlled to be precisely a whole number of wavelengths of 3.39-μm laser radiation coming from a local-oscillator laser, which in turn is frequency-offset locked[1] from the methane-stabilized laser. The frequency of this radiation (and thus the interferometer length) is scanned in the desired manner with a stability nearly equal to that of the methane-stabilized laser. With this technique we have obtained reproducible fringe-pointing precisions of 2×10^{-5} order for the methane line, making it possible to accurately measure the various systematic effects inherent in interferometry.

The measurement system is indicated in Fig. 1. The interferometer was of the plane-parallel Fabry-Perot configuration. The length could be varied from about 1 mm, to give the large free spectral range necessitated by the initially poorly known methane wavelength ($\Delta\lambda/\lambda \approx 10^{-6}$), to a maximum of 25 cm to span the length

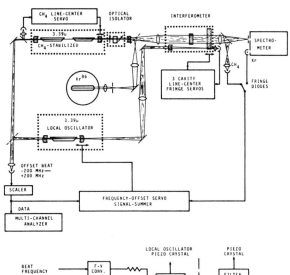

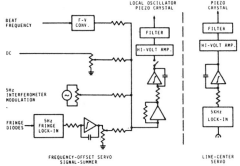

FIG. 1. Frequency-controlled interferometer measurement system.

cluded those due to alignment of light beams not perfectly coaxial and normal to the interferometer, finite size of the pinhole at the Kr fringe pattern center, small plate figure effects, signal-to-noise limitations of Kr, and residual optical feedback effects in the laser case. By using the precision measurement capabilities of the frequency-controlled interferometer, these systematic effects have been carefully measured and corresponding corrections have been made to the wavelength measurement. The total fractional-wavelength uncertainty resulting from these was about 1.8×10^{-9}, with the largest contribution (about 1.5×10^{-9}) coming from misalignment of the Kr light beam (any misalignment gives a unidirectional wavelength shift, but the uncertainty is small enough that no correction has been applied).

Diffraction effects were measured for the 3.39-μm radiation by measuring the effective change ΔT of the interferometer length T as a function of T and interferometer aperture diameter D. Data were least-squares fit to the equation $\Delta T = c(\lambda/D)^2 T$ to determine the diffraction constant $c = 0.056$. For 3.39 μm and for a 1-cm aperture this corresponds to a wavelength correction $\Delta \lambda/\lambda$ of about 6×10^{-9} with an uncertainty of 1×10^{-9}. For comparison, ideal Gaussian modes apertured at both mirrors would give a theoretical value $c = 0.070$.[10]

With the considerations outlined above, we have removed from the data all known inherent systematic effects associated with the interferometric technique to an uncertainty of about 2×10^{-9}. However, the apparent 3.39-μm wavelength varied by $\pm 1.1 \times 10^{-8}$ as a function of the spacer length. This spacer dependence of the effective Kr wavelength is a manifestation of a small intrinsic asymmetry in the Kr emission-line profile and thus is characteristic of all interferometric measurements referenced to the Kr standard. Our observations of the magnitude and sign of this effect are in basic agreement with previous work.[11,12]

Using the precision capability of the frequency-controlled interferometer, it has been possible to carefully measure a radial dependence of the Doppler shift across the Kr discharge capillary,[13] as well as the spacer-dependent change of the effective wavelength due to asymmetry. Indeed, in our experiment the Doppler-shift radial variation was more apparent than is usual,[14] since our technique of lamp observation accepted only light which emerged parallel to the capillary axis and mapped this into the interferometer so that it was parallel to the interferometer axis; this preserved a radial variation of the Doppler shift at the interferometer. In the usual method of observation a light ray at any radius of the interferometer contains light originating at nearly all radii of the discharge; thus all light rays have approximately the same Doppler shift[14] (1.31×10^{-4} cm^{-1}), which is the weighted average over the full discharge bore.

In calculating the necessary Kr wavelength corrections we have assumed that (a) the krypton line asymmetry may be modelled by a doublet[11] and (b) the radially dependent Doppler shift has a smooth form with the intensity-weighted average given by previous studies.[14] The line shape[11] assumed was

(15—20 cm) giving optimum performance with the Kr line. For short cavities, the local-oscillator frequency could not be scanned over a full spectral range of the cavity. Therefore, the motion of the 2-in.-diam scanning flat was controlled with three 30-cm-long servo cavities formed between three spots near the edge of the flat and three small spherical mirrors spaced at 120° around the circumference of the Invar frame. The measurement cavity was formed between the 2-in.-diam flat and a 1½-in.-diam flat which could be translated to the desired interferometer length and adjusted for parallelism inside the vacuum chamber. With the ±250-MHz tuning range of our local oscillator, this allowed the interferometer to be scanned over about one order for methane radiation and nearly six orders for visible radiation. The interferometer length was modulated at 5 Hz, and first-derivative detection was used for fringe pointing. Servo techniques used are indicated in Fig. 1.

Wavelengths have been calculated using the method of fractional fringes. Dispersion of the phase change on reflection was eliminated by subtracting data for two values of interferometer length to obtain data for a virtual spacer. We have used spacers of lengths 4.67, 11.20, 17.71, and 19.56 cm to give the five virtual spacers for which wavelengths have been calculated.

Various systematic offsets were contributed by some of the experimental techniques and by the asymmetry inherent in the Kr line. Effects related to techniques in-

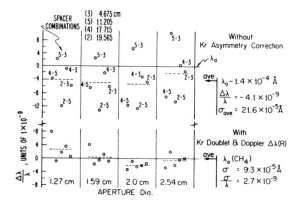

FIG. 2. Wavelength results. Top row corrected only for offsets due to diffraction and finite exit aperture. Bottom row with additional corrections for asymmetry of Kr standard line.

$$I/I_0 = \exp[-\ln 2(\Delta\sigma/\Delta\sigma_{1/2})^2]$$
$$+ 0.06 \exp\{-\ln 2[(\Delta\sigma+\delta)/\Delta\sigma_{1/2}]^2\},$$

where

$$\Delta\sigma = (\sigma - \sigma_0) - \Delta\sigma(r)_{\text{Doppler}}$$

and

$$\Delta\sigma_{1/2} = \text{HWHM} = 0.00634 \text{ cm}^{-1}.$$

After a few iterations, the parameters which approximately fit our data were

$$\delta = 0.008 \text{ cm}^{-1}$$

and, for the discharge capillary of 2.1-mm diameter,

$$\Delta\sigma(r)_{\text{Doppler}} = \Delta\sigma_0 = 2.4\times 10^{-4} \text{ cm}^{-1}, \quad r = 0-0.28 \text{ mm},$$
$$= \Delta\sigma_0 \exp\{-2[(r-0.28 \text{ mm})/0.21 \text{ mm}]^2\},$$
$$r = 0.28 - 1.05 \text{ mm}.$$

Fringe shifts were calculated by folding the asymmetric Kr line shape with the appropriate interferometer fringe-intensity function to give the asymmetric fringe intensity I asym(T) as a function of interferometer length T. Another fringe intensity I sym(T) was calculated for a symmetric Kr line located at some reference point on the asymmetric Kr line (for instance, the maximum intensity).[15] The shift in T of the maximum intensity points of the simulated fringes we denote by ϵ, which typically was a few times $10^{-3}\lambda_{\text{Kr}}/2$.

From the wavelengths uncorrected for asymmetry, a set of ideal ϵ's for the virtual spacers was determined which would entirely remove the apparent CH_4 wavelength variation with interferometer length and aperture. The ϵ's calculated with the above asymmetry model were within about $\frac{1}{2}\times 10^{-3}\lambda_{\text{Kr}}/2$ of these ideal ϵ's for all apertures and spacers. Our CH_4 wavelength variations were reduced by about an order of magnitude by this Kr asymmetry correction.

Wavelengths for all virtual spacers and apertures are shown in Fig. 2, both uncorrected for asymmetry and corrected. Each point represents the wavelength obtained with the indicated virtual spacer (numbers 4–5, etc., next to points in the top row). The two sets in each vertical column are for the indicated aperture. Wavelengths corrected for experimental systematic effects, but not for asymmetry, are shown in the first row (average of all points is indicated by the "ave" arrow at the right). Results for additional corrections due to Kr asymmetry are given in the second row, with the total average λ_0. Averages for each aperture are shown by the dashed lines.

Without asymmetry corrections, the aperture averages deviate from the total average by only about 2×10^{-9}, and the total average is offset from λ_0 by only -4.1×10^{-9}. Despite these small offsets of the wavelength averages, the largest deviations[12] of the λ's is about 1.1×10^{-8} and the small offset of the average λ from λ_0 is only due to the choice of spacers used. This is indicated also by the large standard deviation σ of 21.6×10^{-5} Å ($\sigma/\lambda = 6.4\times 10^{-9}$).

With the inclusion of asymmetry corrections, the largest single deviation is reduced to less than 10^{-8} and the standard deviation to $\sigma = 9.3\times 10^{-5}$ Å ($\sigma/\lambda = 2.7 \times 10^{-9}$). Use of the Doppler $\Delta\lambda(r)$ correction has improved the aperture average λ variation somewhat, but a better hypothetical distribution of velocities in the discharge could probably be found to further reduce this variation.

Our final value λ_0 for the CH_4 wavelength, obtained using the asymmetry discussed above, depends upon how the defined Kr wavelength (6057.802 105 Å) is applied to the line. Since the intrinsic asymmetry of the Kr emission line was discovered only after the meter definition was adopted in 1960, it is necessary to adopt a further convention relating the defined wavelength to the observables of the line. We shall consider two possibilities: (i) The defined wavelength applies to the asymmetric-line maximum-intensity point, or (ii) it applies to the center of gravity point.[16] If the definition corresponds to the maximum intensity, we obtain $\lambda_0 = 33\,922.314\,04$ Å, and, if to the center of gravity, $\lambda_0 = 33\,922.313\,76$ Å.

The uncertainty in the two individual values should probably be larger than the 2.7×10^{-9} standard deviation of the wavelength set due to the uncertainty in the Kr asymmetry model. For instance, if the Rowley-Hamon[11] model is used without the radial Doppler dependence, the wavelength is decreased by 1.2×10^{-9}. Thus, for our result we estimate a 68% confidence interval of

$$\delta\lambda = \pm 1.2\times 10^{-4} \text{ Å},$$
$$\delta\lambda/\lambda = \pm 3.5\times 10^{-9}$$

by combining quadratically the 2.7×10^{-9} standard deviation with two times the 1.2×10^{-9} model dependence shift.

This methane wavelength has also been measured by Giacomo.[17] Wavelengths inferred from his measurement for the two cases discussed here are somewhat lower than ours, although his quoted result (33 922.313 76 Å) is identical to ours for our case where the defined Kr wavelength is applied to the line center of gravity.

Using the measured methane-stabilized laser frequency[5] of $\nu = 88.376\,181\,627(50)$ THz ($\delta\nu/\nu = \pm 5.6\times 10^{-10}$) and the presently reported wavelength of $\lambda = 3.392\,231\,376(12)$

µm ($\delta\lambda/\lambda = \pm 3.5 \times 10^{-9}$), we calculate the speed of light to be $c = 299\,792\,456.2\,(1.1)$ m/sec ($\delta c/c = \pm 3.6 \times 10^{-9}$). This value is based on the arbitrary convention that the krypton defined wavelength is to be applied to the center of gravity of the krypton line; if the maximum intensity point were chosen instead,[16] the methane wavelength—and hence the value of c—would be increased by 8.3 parts in 10^9 [$c_{\text{max I}} = 299\,792\,458.7\,(1.1)$ m/sec]. These values are in agreement with the previously accepted value[18] of 299 792 500 (100) m/sec and are about 100 times more accurate. They are also in agreement with the recent measurement by Bay, Luther, and White,[19] who report $c = 299\,792\,462\,(18)$ m/sec.

The authors would like to express their thanks to Dr. P. Bender and Dr. H.S. Boyne of these laboratories for their helpful discussions during the course of this work. Also, the authors extend special appreciation to P. Giacomo of BIPM for his very helpful criticisms and comments.

---

*Operated jointly by the National Bureau of Standards and the University of Colorado.

[1] R.L. Barger and J.L. Hall, Phys. Rev. Lett. 22, 4 (1969).
[2] E.E. Uzgiris, J.L. Hall, and R.L. Barger, Phys. Rev. Lett. 26, 289 (1971).
[3] K. Uehara, K. Sakurai, and K. Shimoda, J. Phys. Soc. Jap. 26, 1018 (1969).
[4] J.L. Hall in *Proceedings of the Symposium on Basic and Applied Laser Physics*, Esfahan 1971 (Wiley, New York, to be published).
[5] K.M. Evenson, J.S. Wells, F.R. Petersen, B.L. Danielson, and A.W. Day, preceding letter, Appl. Phys. Lett. 22, 192 (1972).
[6] K.M. Evenson, J.S. Wells, F.R. Petersen, B.L. Danielson, G.W. Day, R.L. Barger, and J.L. Hall, Phys. Rev. Lett. 29, 1346 (1972).
[7] E. Engelhard and J. Terrien, Rev. Opt. 39, 11 (1960).
[8] This lamp was made available by Dr. G. Schweitzer and Dr. R. Deslattes of Natl. Bur. Std., Gaithersburg, Md.
[9] R.L. Barger and J.L. Hall in *Precision Measurement and Fundamental Constants*, Natl. Bur. Std. Special Publication No. 343 (U.S. GPO, Washington, D.C., 1971).
[10] Calculated from Fig. 24 of H. Kogelnik and T. Li, Appl. Opt. 5, 1550 (1966). See also L.A. Vainshtein, Zh. Eksp. Teor. Fiz. 44, 1050 (1963) [Sov. Phys. JETP 17, 709 (1963)].
[11] W.R.C. Rowley and J. Hamon, Rev. Opt. 42, 519 (1963).
[12] This Kr^{86} reproducibility limit is just larger than the 1×10^{-8} stated in the report of the Comité Consultatif pour la definition du Metre (1970) (Bureau International des Poids et Mesures Sevres, France, 1972).
[13] A radial variation of the Doppler shift has been postulated by F. Bayer-Helms [Z. Angew Phys. 15, 330 (1963); 15, 416 (1963); 16, 44 (1963)].
[14] K.M. Baird and D.S. Smith, Can. J. Phys. 37, 832 (1957); see also Ref. 11.
[15] For a more detailed discussion of asymmetry corrections, see R.L. Barger and J.L. Hall (unpublished).
[16] In the absence of an international agreement on this question, we slightly prefer the center of gravity definition since it probably would be less affected by lamp operating conditions (influence on Doppler width).
[17] P. Giacomo in *Proceedings of Fourth International Conference on Atomic Masses and Fundamental Constants*, Teddington, 1971 (unpublished).
[18] K.D. Froome, Proc. Roy. Soc. Lond. 247A, 109 (1958).
[19] Z. Bay, G.G. Luther, and J.A. White, Phys. Rev. Lett. 29, 189 (1972).

LOCKING A LASER FREQUENCY TO THE TIME STANDARD

Z. Bay and G. G. Luther
National Bureau of Standards,
Washington, D. C. 20234
(Received 5 August 1968)

On the basis of high-frequency modulation experiments and well-known locking techniques, a scheme is described for stabilizing a visible laser frequency and simultaneously determining that frequency in terms of the time standard. The importance of this method for a refined determination of the velocity of light and the possibility of establishment of reference lines for spectroscopy and for length measurements, throughout the spectrum wherever laser lines are available, is discussed.

We have demonstrated in our recent experiments[1] that an electrooptic intracavity amplitude modulation of a 6328-Å He–Ne laser is practicable at frequencies, ω, up to 25 GHz. The yield in the side bands, $\nu \pm \omega$ (where ν is the laser frequency), exceeded 10^{11} photons per second. There were no indications in the experiments that these numbers should be considered as upper limits. These results, along with the ability demonstrated by many experimenters[2-5] to lock a laser to a passive reference cavity assures one of the feasibility of locking the frequency of a visible laser line to a microwave frequency, and of relating thereby that optical frequency to the time standard. The method for accomplishing this as being pursued in this laboratory is as follows.

The difference frequency between the sidebands is known (2ω), and the ratio $(\nu + \omega)/(\nu - \omega)$ can be established as corresponding to the ratio N_+/N_- of the two order numbers N_+ and N_- of a passive cavity, when that cavity is tuned simultaneously to both frequencies.[6] Then, using the first approximation theory of Fabry–Perot cavities,

$$\nu + \omega = N_+(c/2L)$$
$$\nu - \omega = N_-(c/2L) \,, \quad (1)$$

where c is the velocity of light and L is the length of the interferometer cavity. Thus,

$$\nu = 2\omega(N_+/n) - \omega = 2\omega(N_-/n) + \omega \,, \quad (2)$$

where $n = N_+ - N_-$.

Hence, ν is established relative to ω and is known if the integer order numbers N_+ and N_- are known. It is seen from Eq. (2) that neither the knowledge of the velocity of light nor that of the length of the cavity is required in the measurement of ν, even though both are important parameters in the operation of the cavity.

In order to realize by servo mechanisms the two conditions as expressed by Eqs. (1), two error signals are needed. At a preset value of ω, L and ν must be driven to the appropriate values. The two error signals can be derived by sweeping ω through $\beta \cos 2\pi\gamma_\omega t$ and simultaneously sweeping L through $b \cos 2\pi\gamma_L t$. Both γ_ω and γ_L are low frequencies with respect to the bandwidth of the interferometer (in our apparatus $\gamma_\omega = 2$ kHz and $\gamma_L \simeq 50$ kHz). The amplitudes β and b are chosen to be about the half width of the Fabry–Perot transparency function.

The ω sweep moves the side-band frequencies in opposite directions. Thus the contributions to the error signal (the transmitted light intensity, phase detected at γ_ω) will be of opposite sign for the two side bands. When the length L is appropriate to ω these contributions cancel and the error signal is zero. This error signal is used to servo L to ω.

The error signal phase detected at γ_L is used to servo the length of the laser cavity, and consequently ν, to L. It should be emphasized that the error signals detected at the frequencies γ_L and γ_ω, though coming from the same photodetector, are independent if these two frequencies are unrelated. Thus the two servo systems work simultaneously (Fig. 1).

The error signal detected at γ_ω is highly insensitive to fluctuations in ν, because this signal is the sum of the contributions of opposite sign from the two side bands. Each of those components may undergo significant fluctuations (due to fluctuations in ν) but those fluctuations are strictly correlated and cancel in the algebraic sum. The only uncorrelated part is that due to shot noise. Therefore a long integration time can be used to overcome shot noise limitations. With photoelectron currents of 10^{10} electrons per second in each side band and a one second integration time (relative shot noise fluctuation $\sim 10^{-5}$) the setting of the interferometer length to $\delta L/L \sim 10^{-5} \Delta\nu/2\omega \sim 10^{-9}$ is expected if the finesse $F \sim 100$, L is of the order of one meter ($\Delta\nu = c/2LF \sim$ 1.5 MHz), and $\omega \sim 10$ GHz. This corresponds to setting L to 1/3 of the fringe-width (~ 10 Å).

Thus, the simultaneous operation of the two servo systems readily assures setting of the optical frequency, ν, to 1 part in 10^9 (~ 500 kHz).

A more exact statement of Eq. (2) which does not neglect diffraction and reflection phase shifts is

$$\nu = 2\omega[N_+/(n + \mu)] - \omega \,. \quad (2a)$$

The μ which incorporates the effect of phase shifts can be determined to the necessary accuracy (sufficient to utilize the achieved accuracy of L) by well-known techniques of interferometry, that is, by using a fictitious cavity length which is the difference in length between two successive settings of the interferometer.

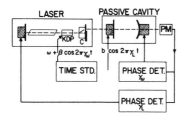

Fig. 1. Locking a laser frequency to a frequency standard. The 6382-Å He–Ne laser is modulated electrooptically by an intracavity KDP crystal driven at the microwave frequency ω; the side bands $\nu \pm \omega$ are coupled out by a calcite plate C, their polarization being opposite to that of the main laser beam; ω is locked to a frequency standard and is also swept at a low frequency γ_ω. The side bands pass through the passive cavity and are detected in the photomultiplier, PM. The length of the passive cavity L is modulated at the frequency γ_L. The output of the PM, as phase-detected at γ_ω servoes L, and as phase-detected at γ_L, simultaneously servoes the laser. The result is that the passive cavity is tuned to both side bands and ν is stabilized to ω.

The precision of the above method may be increased by increasing (1) the length and/or the finesse of the interferometer; (2) the modulation frequency; (3) the intensity of the side bands; (4) the integration time of the length servo. The ultimate long term stability is limited only by that of the time standard.

It should be emphasized that, while consisting of components performing similar operations, the above scheme differs from those of Refs. 2–5 in the sense that in those experiments the passive cavity is locked to atomic or molecular resonances, individually chosen for each laser line. In the present scheme one single resonance transition is used for any ν, namely, that of the frequency standard. This automatically relates every ν to the internationally accepted frequency scale. Thus direct frequency measurements which have already been achieved in the far infrared[7,8] may be compared with future measurements made in the visible without the difficulties inherent in the intercomparison of wavelengths in distant parts of the spectrum.

The above method of stabilization and frequency measurement is applicable to any laser line. Also, the frequency of laser lines, stabilized by other means, can be measured by these techniques. Thus, as one important application of these ideas, reference lines for spectroscopy can be established throughout the spectrum wherever laser lines are available.

Another application is the evaluation of the velocity of light. It is obvious that the knowledge of the frequency in terms of the standard second and a simultaneous measurement of the corresponding wavelength in terms of the length standard results in refining the value of the velocity of light, c, in meters/second. Since it is expected that the accuracy of the ν measurement will surpass that of the definition of the present length standard, the accuracy of c will be limited to that of the present meter, about 1 part in 10^8. A definition of c (compatible with the present meter, but otherwise arbitrary) would result in a new definition of the meter. As the above method is applicable to any laser line, reference wavelengths for precise length measurements would be available in any part of the optical spectrum without the need to define one particular wavelength as a new length standard. Also, time of flight measurements (in terrestrial and space radar) could be translated directly into distances via the definition of c.

---

[1] Details to be published later. Our experiments extend the work of R. Targ, G. A. Massey, and S. E. Harris, Proc. IEEE, 52, 1247 (1964), to higher modulation frequencies.
[2] A. D. White, E. I. Gordon, and E. F. Labuda, Appl. Phys. Letters 5, 97 (1964).
[3] A. D. White, IEEE J. Quantum Electron. QE-1, 322 (1965).
[4] A. D. White, Rev. Sci. Instr. 38, 1079 (1967).
[5] S. Ezekiel and R. Weiss, Bull. Intern. Quantum Electron. Conference, Miami, 1968, p. 53.
[6] Z. Bay and G. G. Luther, Natl. Bur. Std. Tech. Note 449, 30 (1968).
[7] L. O. Hocker, A. Javan, D. Ramachandra Rao, L. Frenkel, and T. Sullivan, Appl. Phys. Letters 10, 147 (1967).
[8] L. Frenkel, T. Sullivan, M. A. Pollack, and T. J. Bridges, Appl. Phys. Letters 11, 344 (1967).

Description, Performance, and Wavelengths of Iodine Stabilized Lasers

W. G. Schweitzer, Jr., E. G. Kessler, Jr., R. D. Deslattes, H. P. Layer, and J. R. Whetstone

A description is given of lasers stabilized to components of the $^{129}I_2$ spectrum in the region of the 633-nm laser lines for ^{3}He–^{20}Ne and ^{3}He–^{22}Ne. Relationships between operational characteristics such as power output, peak size, and peak width are shown, along with their relationships to some of the controllable parameters such as excitation level, iodine absorption, and iodine pressure. We found an iodine pressure broadening of about 13 MHz/Torr with a 2.6-MHz zero-pressure intercept. The frequency shift associated with iodine pressure is roughly 2×10^{-9} ν/Torr to the red. Power broadening and power shifts are small, about a 10% increase in width and about 2×10^{-11} ν variation in frequency for a fivefold to sixfold increase in power. These lasers exhibit a frequency stability for 10-sec sampling time of about 2×10^{-12} ν and a resetability of about 1×10^{-10} ν. The absolute vacuum wavelength for one iodine component has been measured against the ^{86}Kr standard—^{3}He–^{20}Ne:$^{129}I_2$, k λ = 632 991.2670 ± 0.0009 pm. The wavelengths of several other iodine components have been determined by measuring the frequency difference between them and the $^{129}I_2$, k component. Among these are ^{3}He–^{22}Ne:$^{129}I_2$, B λ = 632 990.0742 ± 0.0009 pm; and ^{3}He–^{20}Ne:$^{127}I_2$, i λ = 632 991.3954 ± 0.0009 pm. These results were obtained using the Rowley-Hamon model for asymmetry in the krypton line and assume that the defined value for the standard is associated with the center of gravity of the line profile. The indicated uncertainties are statistical. No allowance has been included for imperfect realization of the krypton standard or for uncertainty in the asymmetry model.

Introduction

The powerful technique of utilizing saturated absorption to eliminate Doppler broadening was first demonstrated by Lee and Skolnick.[1] Since that time it has been used with a number of molecular species for laser stabilization and high resolution spectroscopy. Barger and Hall[2] applied this technique to stabilize a He–Ne laser to an absorption line in methane at 3.39 µm. Hanes and Dahlstrom,[3] Wallard,[4] and Hanes et al.,[5] applied it to stabilize a ^{3}He–^{20}Ne laser to hyperfine components in the absorption spectrum of $^{127}I_2$ at 633 nm, and Knox and Pao[6] to stabilize a ^{3}He–^{20}Ne laser to hyperfine components in the absorption spectrum of $^{129}I_2$ at 633 nm.

Because of their promise as stable wavelength and frequency sources we set out to study the performance of iodine stabilized lasers at 633 nm, to make absolute wavelength measurements of some of the more promising lines, and to compare them with methane stabilized lasers at 3.39 µm. Another paper[7] describes a high resolution interferometric comparison of an iodine stabilized laser with a methane stabilized laser. In this paper we will give the results of our studies of several embodiments of iodine stabilized lasers, operational details, stability, reproducibility, and absolute wavelength determinations.

Iodine Stabilized Lasers

Figure 1 is a cutaway view of one of these lasers. (Not shown are enclosures for the air spaces between elements in the cavity.) The general arrangement is similar to other such schemes with intracavity absorbers. The mirrors are mounted in holders at least one of which also contains a piezoelectric transducer. The holders are fastened to the end plates with push-pull screws, which provide a convenient range of adjustment and adequate rigidity. The spacer rods are Invar. Cavity length ranges from about 20 cm to 30 cm depending on the lengths of the plasma tube and absorption cell. The laser tube and the absorption cell are held onto the rods in kinematical mounts. The holder for the absorption cell also contains a thermoelectric cooler for controlling the temperature of the cell cold finger. The body of the holder also serves as a heat sink for the warm junction of the cooler.

Figure 2 is a block diagram of the stabilization

The authors are with the National Bureau of Standards, Washington, D.C. 20234.
Received 15 June 1973.

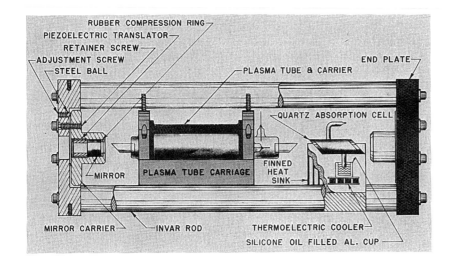

Fig. 1. Cutaway view of I_2 laser.

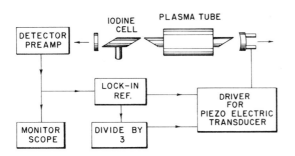

Fig. 2. Block diagram—stabilization scheme.

scheme. The electronic part of the servo loop consists of a silicon photodiode and current-to-voltage converter, a lock-in amplifier, and a driver for the piezoelectric transducer. The reference input to the driver provides for the small modulation of the laser length required for the lock. The divide by three circuit is used when a third derivative lock is desired.[4,5] It is by-passed for a first derivative lock.

The shape of the tuning curve, power output vs cavity length from one order to the next, is determined by the combined effects of the laser tube output (including Lamb dip), background absorption of iodine lines within the Doppler range, and other cavity losses. The slope under a particular iodine peak will therefore depend on all of these parameters in addition to its wavelength relative to the 633-nm laser line. We have looked for and found conditions that result in the desired peak being on a maximum of the power tuning curve. Under these conditions, one can make use of a first derivative, peak-seeking servo system for stabilization. To eliminate small errors due to imperfect realization of this condition we have gone to the third derivative locking scheme, which is illustrated in Fig. 2.

Most of our work has been done with $^{129}I_2$, using either ^{20}Ne or ^{22}Ne in the plasma tube. We have also constructed an $^{127}I_2$ stabilized laser similar to that described by Hanes et al.[5] in order to effect one of the wavelength comparisons described below.

Iodine and Cells

Iodine-129 is very slightly radioactive but not at all difficult to handle on this account.[8] It was more of a problem that the iodine crystals supplied to us were contaminated with some of the solvent (acetone in this case) used in separating the element after its reduction from the salt. We were able to remove this satisfactorily by differential trapping and pumping under vacuum, using several cold fingers and dif-

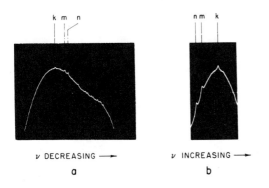 

Fig. 3. (a) Tuning curve for ^{3}He-^{20}Ne:$^{129}I_2$ system. Peak k on top. (b) High dispersion view of (a). (c) Still higher dispersion view of (a).

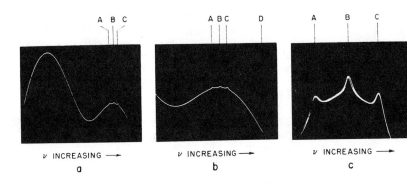

Fig. 4. (a) Tuning curve for ^{3}He–^{22}Ne:^{129}I$_2$ system. Peak B on top. (b) Higher dispersion view of (a) with peak D just showing. (c) Still higher dispersion view of (a).

ferent cooling baths. In order to separate out a possible contamination by water we passed the I$_2$ vapor over P$_2$O$_5$ during the process of purifying and transferring to the ampules from which we fill the cells.

The cells are quartz with quartz windows fused on. We tried epoxy seals first, but these are quickly attacked by iodine. We also tried Mylar[9] seals (quartz cells and windows). These were much better than epoxy but also failed after a few months due to an apparent reaction between the iodine and the seal.

We have used cell lengths from 1.3 cm to 10 cm and cold finger temperatures ranging between room temperature and −15°C. This corresponds to a vapor pressure range of about 6 mTorr to 200 mTorr.[10] The cold finger temperature is conveniently maintained by a small thermoelectric cooler.

^{3}He–^{20}Ne:^{129}I$_2$ System

This is the system described by Knox and Pao[6] and is the one we tried first.[11] A tuning curve for this case is shown in Fig. 3, a, b, and c. There are a variety of combinations of plasma tube fill, excitation level, and iodine absorption, which will result in k being on top as shown. In this particular case the plasma tube was filled to 1.85 Torr of 7:1 ^{3}He–^{20}Ne and had a 2-mm bore with 18-cm active length and 4-mA discharge current. The iodine cell was 3 cm long, and the cold finger was at 0°C. Under different conditions (lower excitation and/or lower absorption) m or n will lie at or near the top. (m is a superposition of at least two components.) It is, of course, easy to lock on any of the resolved peaks using the third derivative control system, but we prefer to choose a peak on a tuning curve maximum if we can.

^{3}He–^{22}Ne:^{129}I$_2$ System

This system has not been described before. It has a number of desirable properties that make it more attractive than the other system. A tuning curve for this system is shown in Fig. 4, a, b, and c. In this case the plasma tube was filled to 1.85 Torr of 7:1 ^{3}He–^{22}Ne and had a 2-mm bore with 18-cm active length and 6-mA discharge current. The iodine cell was 3 cm long, and cold finger was at 10°C. We emphasize again that these are not the only conditions that result in the triplet falling on a maximum of the tuning curve. As will be seen below, most of our data on these peaks were obtained with a shorter plasma tube at slightly higher pressure and lower current. The Lamb dip is slightly shallower in that case and disappears if the plasma tube pressure is raised much higher than 2.5 Torr (depending on the iodine absorption, of course).

In Fig. 4, starting with the conspicuous triplet and going to higher frequencies, we have labeled these peaks A, B, C, and D (which is just visible at the right edge of Fig. 4b). There are at least five other peaks beyond D, toward higher frequencies that we have observed with a shorter cavity and that appear to belong to this group of components. Not apparent in the photographs of Fig. 4 but observable at higher amplification and with more absorption is a group of at least twenty-seven smaller components distributed almost uniformly over the low frequency half of the laser line. We have mapped out these components, with a third derivative detector. Their positions and relative sizes are shown in Fig. 5, along with A, B, C, D, and the approximate location of the Lamb dip of the laser line.

The triplet A, B, C is well separated from other components and, as will be shown below, appears to have a saturation parameter intermediate between that for the ^{3}He–^{20}Ne:^{129}I$_2$ system and the ^{3}He–^{20}Ne:^{127}I$_2$ system. The range of useful operating conditions is much greater for A, B, C than for the other alternatives.

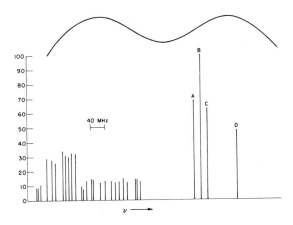

Fig. 5. Map of ^{3}He–^{22}Ne:^{129}I$_2$ system.

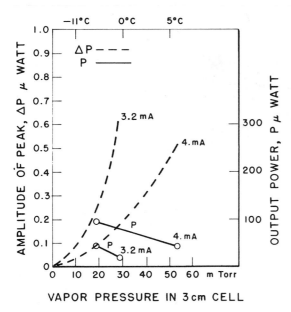

Fig. 6. Power output, P, and absorption peak size, ΔP, as a function of iodine absorption for various levels of excitation (laser current). ^{3}He–^{20}Ne:^{129}I$_2$ k peak.

Operating Conditions

We wished to study the characteristics of these iodine stabilized lasers over as large a range of operating conditions as feasible. However, in order to keep the effort within reasonable bounds we fixed some of the variable parameters at values that prior experience had indicated were most useful. Thus, the laser cavity was about 26 cm long and consisted of a 2-m radius 1.2% T mirror at one end and a 0.6% T flat at the other end. The iodine absorption cell was 3 cm long. The over-all length of the plasma tube was 16 cm, and the active discharge length was 10.5 cm. The ^{20}Ne tube had a 2-mm capillary and was filled to 1.85 Torr of 7:1 ^{3}He–^{20}Ne. The ^{22}Ne tube had a 1.5-mm bore and was filled to 2 Torr of 6:1 ^{3}He–^{22}Ne.

Spectral data were taken by applying a slow linear ramp and a small 600-Hz modulation to the cavity length and recording the output of a phase sensitive detector tuned to the third harmonic of the modulation. The modulation amplitude, kept constant for all runs, was made small enough that it contributed little to the width of the peaks. The third derivative signals were much smaller under these conditions than they are for optimum third derivative signal.

Power Output and Peak Size

Figures 6 and 7 show the power output, P, and absorption peak size, ΔP, as a function of iodine absorption for various levels of excitation (laser current). The cavity losses, excluding iodine absorption, are the same for both figures and are largely due to the total of about 1.8% per pass transmitted by the mirrors. We have not measured the absorption coefficient for ^{129}I$_2$ at these two wavelengths,

but from Hanes et al.[12] and from Knox and Pao[13] we estimate α to be about 2 per meter-Torr (transmittance $T = \exp(-\alpha pl)$. For the 3-cm cell, this would lead to a single pass absorptance of 0.15% for a vapor pressure of 25 mTorr and proportionately more or less at other pressures.

These curves were derived from a large number of curves similar to that in Fig. 8a. The range of data acquired for the ^{20}Ne–^{129}I$_2$ k peak was limited at higher powers by the reduced size of the peak and at low powers by hysteresis.[14] Hysteresis is usually conspicuous in tuning curves for the ^{20}Ne–^{129}I$_2$ system but does not present a problem when locking to the k peak inside the range of parameters indicated. The small modulation in cavity length required for locking is not sufficient to take it into the region where the laser can suddenly switch to the zero power state.

For the ^{22}Ne–^{129}I$_2$ B peak the range of conditions yielding useful output powers and good absorption peaks was much greater. Hysteresis was not evident except at the highest power levels with high iodine absorption. Even there it was a much smaller effect than that observed for the k peak (which is associated with ^{20}Ne emission).[15] We have not been able to calculate the saturation parameter satisfactorily for the iodine absorber. We can make some estimates based on data such as that in Fig. 8a for the k peak. The size of the peak is reduced by one half in going from 20-μW to 40-μW output. We could not get data below 20 μW because of hysteresis. At 40-μW output the power density inside the iodine cell (beam area about 4×10^{-3} cm^2) is approximately 1 W/cm^2. The saturation parameter for the gain tube at our gas pressure is given by Smith[16] as about 15 W/cm^2. It would appear that the ratio of the two saturation parameters gain/absorption is at least 15 and perhaps even greater at this iodine pressure (19 mTorr). We did not carry similar curves for the ^{22}Ne–^{129}I$_2$ B peak below about 50 μW. We believe that the saturation parameter for this case is at least several times as great as that for the k peak.

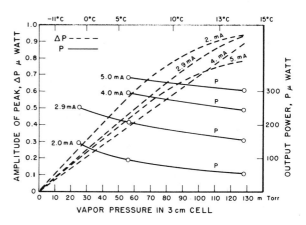

Fig. 7. Power output, P, and absorption peak size, ΔP, as a function of iodine absorption for various levels of excitation (laser current). ^{3}He–^{22}Ne:^{129}I$_2$ B peak.

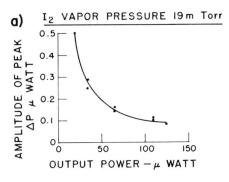

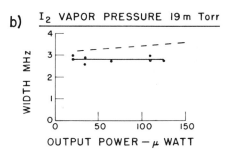

Fig. 8. (a) Absorption peak size ΔP and (b) absorption peak width, as a function of output power through 1.2% T mirror. Iodine vapor pressure is about 19 mTorr. Dashed curve in (b) is at a slightly higher modulation level. ^{20}Ne:^{129}I$_2$ k peak.

Pressure Broadening and Power Broadening

We measured the widths of the peaks from the third derivative recordings. There was quite a bit of scatter in these because the laser was not stabilized during the relatively long scan period and because we kept the modulation low to contribute as little as possible to the width. In spite of this we were able to establish two facts reasonably well. The power broadening is not very large, and there is a significant pressure broadening. In Fig. 8b note that for the usual modulation level we could not really discern any broadening in the k peak as we increased the power output from 15 μW to 125 μW. At a slightly greater modulation level the scatter is reduced somewhat, and we recorded the trend shown in the dashed curve, which shows an increase in width from ~3.2 MHz at 25 μW to about 3.6 MHz at 150 μW. This is only about 10% increase in width for a sixfold increase in power. Results for the B peak were about the same. This result is for a single iodine vapor pressure. When we repeated the experiment at different iodine pressures we found that there was a pressure broadening of about 2.4 MHz for 180 mTorr or 13 MHz/Torr with a zero pressure intercept of ~2.6 MHz. This is shown in Fig. 9, where both the k peak and the B peak data fall nearly on the same line. All the data at each pressure were averaged together so that the error bars include whatever power broadening there may be along with the scatter.

The lines saturate strongly and yet have relatively little power broadening compared with the observed width. We can only speculate that there is another mechanism for broadening, besides pressure and power, that dominates at low pressure.

Stability and Resetability

The two lasers used to study stability and resetability were similar in construction and operating characteristics to those described above. At the time we had only one ^{22}Ne plasma tube available (two would have been required) so this study was carried out using ^{3}He–^{20}Ne:^{129}I$_2$ devices. Both the lasers were similar to those described under *Operating Conditions* but had 1.3-cm absorption cells. One was locked to the n peak and the other to the k peak. Stability and reproducibility was studied by examining the beat signal produced by heterodyning these two lasers. The laser stabilized to n was run at constant power and iodine cell temperature, while that stabilized to k had its operating parameters varied in a systematic manner so that variations in the approximately 55-MHz beat signal could be attributed to its frequency alone.

The experimental arrangement utilizing two third harmonic locks is shown in Fig. 10. One oscillator was used to control both lock-in stabilizers; the laser frequencies were made to track by adjustment of the phase and amplitude of the modulating signals used to drive the piezoelectric mirror mounts. The residual frequency modulation of the beat signal was masked by statistical effects in the visual display of the multichannel scaler for all but the shortest sample times (2×10^{-4} sec). To minimize environmental disturbances, the optical apparatus was mounted on an air suspension bench and was located remotely from the electronic control hardware.

A stability index for the lasers was derived from the sequential observation of the beat frequency for

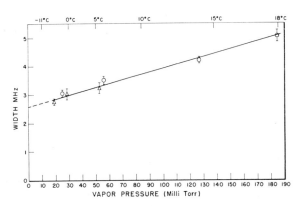

Fig. 9. Absorption peak width as a function of iodine vapor pressure. Cold finger temperature indicated for convenience at several points on pressure scale. ○, ^{22}Ne:^{129}I$_2$ B peak; △, ^{20}Ne:^{129}I$_2$ k peak.

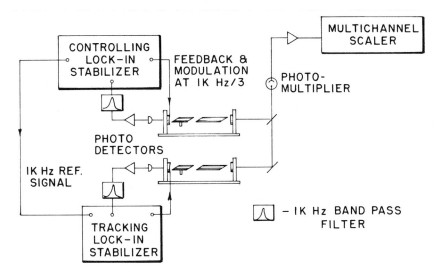

Fig. 10. Experimental arrangement for stability measurements.

various sampling intervals. From such observations we calculated

$$\sigma_{\Delta\nu/\nu}(2,\tau,\tau) = \frac{1}{4.73 \times 10^{14}} \left[\frac{1}{2} \frac{\sum_{i=1}^{N}(\nu_i - \nu_{i+1})^2}{2(N-1)} \right]^{1/2},$$

where i ranges over successive measurements of the average frequency during the time interval τ (no intervening dead time) This expression is the square root of half of the two-sample Allan variance.[17] The factor of 1/2 assumes an equal contribution from each laser, which is justified by the formal similarity of their operating characteristics. The results of these measurements are shown in Fig. 11 for sample intervals between 2×10^{-4} sec and 200 sec. The data points for time intervals below 10 sec were calculated from 250 consecutive sample intervals, while those above 10 sec were calculated from 25 in order to reduce measurement time. The uncertainty in $\sigma_{\Delta\nu/\nu}$ was estimated in the following way: a single block of data containing 250 separate measurements was converted into 25 blocks containing 10 measurements apiece. The error bars in Fig. 11 represent the *a posteriori* standard deviation of the 25 values of $\sigma_{\Delta\nu/\nu}$ calculated from each of the 25 groups of 10 measurements. The highest stability is found above the 10-sec interval and is about $2 \times 10^{-12} \nu$. The stability for smaller time intervals is limited by electronic control circuitry and for longer intervals by longer period environmental changes; and there is reason to believe that this performance could be improved. The results in Fig. 11 were obtained under favorable operating conditions for the ^{20}Ne-^{129}I$_2$ system. The absorption cell pressure of 130 mTorr (giving the same absorptance as 56 mTorr in the 3-cm cell) provided a relatively large absorption peak at a relatively low excitation level symmetrically situated on a background maximum.

The wideband SNR in this case was about 20 dB and, in general, is dominated by the noise characteristics of the gain tube. The ability to relocate the lock point electronically by breaking and reacquisition (electronic resetability) was $3 \times 10^{-12} \nu$ for a 5-sec measurement interval, which is consistent with the Allan variance data.

Measurement of reproducibility of independent lasers and of dependence of laser frequency on operating parameters such as power level, iodine absorption cell pressure, other cavity losses, and background slope are difficult because of the manner in which they are interrelated. It is nevertheless important to estimate their independent effects as far as possible.

By judicious choice of power, cell temperature, and internal losses (Brewster's angle mismatch between gain tube and absorption cell) the effect of these parameters on frequency can be sorted out to some extent. The effect of power level was studied by operating the cell at 100 mTorr and adjusting the cavity losses so that changes in power level produced only a slight change in the background slope under the absorption peak. In this manner the power level in the cavity was varied over a factor of 6 with concomitant variations in frequency of $2 \times 10^{-11} \nu$, an order of magnitude greater than the electronic

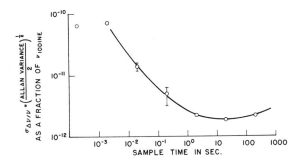

Fig. 11. Allan variance plots of iodine stabilized laser frequency stability.

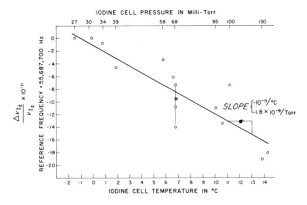

Fig. 12. Frequency shift due to iodine pressure.

resetability limit. When operating at 60 mTorr, however, a variation in power level shifts the background slope quite readily so that a change in sign of the slope can be effected. Shifting the background maximum from one side of the absorption peak to the other (a relative shift of background maximum of 20 MHz relative to the absorption peak maximum) produces a frequency shift of $1.2 \times 10^{-11} \nu$, again about an order of magnitude greater than the electronic resetability limit. It is worth mentioning, in passing, that the third harmonic lock discriminates against the effects of background slope much more effectively than does the fundamental lock. When the above measurement was repeated using a fundamental lock, the frequency shifts associated with changes in the background slope were two orders of magnitude greater than those experienced with the third harmonic lock.

Third derivative locking schemes do have problems that require some caution; the third harmonic error signal has three components: The dominant signal generated from the fundamental scan and the absorption peak, a smaller signal generated from the fundamental scan and the background slope, and an unwanted signal generated by spurious third harmonic content in the modulating signal and/or nonlinearity in the piezoelectric mirror translator transfer characteristics. In our measurements, when stabilized lasers were operated as outlined above, with the absorption peak near the zero slope portion of the background, the effect of nonlinearity and spurious third harmonic in the modulation were no larger than $1 \times 10^{-11} \nu$.

The effect of changing cell temperature and, consequently, $^{129}I_2$ pressure is shown in Fig. 12. The magnitude of this effect over the operating range of the stabilized laser is about $2 \times 10^{-10} \nu$. The slopes of the relative frequency changes, if linear relationships are assumed, are roughly $-1.8 \times 10^{-9} \nu/$Torr and $-1 \times 10^{-11} \nu/°$C. The scatter of these data is quite large because it was necessary to vary internal losses by misaligning the gain and absorption cell Brewster's angle windows in going from high to low pressure in a manner that is difficult to characterize quantitatively. This procedure was used in order to keep the absorption peak near the maximum of the tuning curve. The data points at 68 mTorr, which are connected by a straight line, illustrate a limitation on resetability. By just such an adjustment a spread of $6 \times 10^{-11} \nu$ was observed and represents the largest uncertainty in the $^{129}I_2$ stabilization system at this time. As indicated above, power shifts are not large enough by themselves to explain this large a shift. Although this spread remains, at present, a limitation on resetability that we do not completely understand, our experience indicates that the resetability of the lock point frequency for lasers operating at the same iodine pressure but with various reasonable combinations of the other operating conditions is about $1 \times 10^{-10} \nu$.

Wavelength Measurements

^{3}He–^{20}Ne:$^{129}I_2\ k$

The absolute wavelength of the 633-nm ^{3}He–^{20}Ne laser stabilized to the $^{129}I_2\ k$ component was measured by interferometric comparison with the ^{86}Kr 606-nm primary standard line.[18] A pressure-scanned flat-plate Fabry-Perot interferometer with photoelectric detection was employed to make this comparison and is shown schematically in Fig. 13.

Light from the iodine stabilized laser was diverged by a strong negative lens, L_1, before being incident on the rotating ground glass screen, S. A first derivative lock was used to stabilize the laser, and the background under the k peak was adjusted for zero

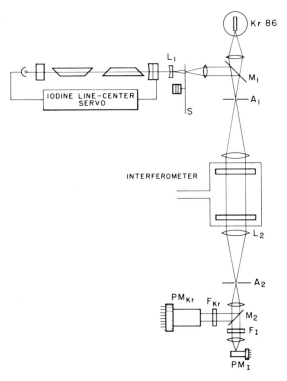

Fig. 13. Experimental arrangement for absolute wavelength measurements.

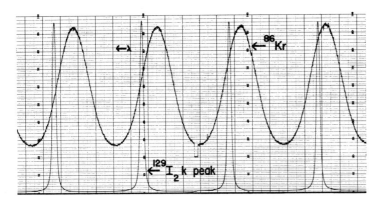

Fig. 14. Typical strip chart record of Fabry-Perot fringes obtained during wavelength measurement. The spacer length was 162 mm.

slope. This procedure though cruder than those described earlier is appropriate to the limitations of the ^{86}Kr source. The light scattered from S was combined with the ^{86}Kr radiation by means of a beam splitter M_1, and both light beams were focused onto A_1 the entrance aperture for the interferometer. The ^{86}Kr lamp was constructed at PTB #76586) and operated in accordance with the recommendations of the International Committee of Weights and Measure (CIPM)[19]—the inside diameter of the capillary was 2.04 mm, the current was 10.8 mA (current density 0.33 A/cm²), the capillary was immersed in a refrigerating bath held at the triple point of liquid nitrogen, and the lamp was viewed end-on with the anode toward the interferometer. The CIPM recommended current density is 0.30 ± 0.1 A/cm². A systematic error[20,21] of ~3 parts in 10^{10} results as the current density is changed from 0.30 A/cm² to 0.33 A/cm². Because this error is both small and uncertain it has not been included in our data. The collimated light beam passed through the interferometer, and the resulting fringes were focused onto the interferometer exit aperture A_2. A second beam splitter M_2, two narrow pass interference filters F_{Kr} and F_I, and two photomultipliers PM_{Kr} and PM_I provided a means to simultaneously record the two sets of fringes. Great care was taken to ensure uniform illumination of aperture A_1 and the etalon plates by both light sources. The accurate positioning of aperture A_2 at the center of the fringe systems was accomplished by use of a microscope.

The etalon consisted of 10-cm diam dielectrically coated plates of which only the central 5-cm diam was used for the measurements. The plates were separated by Invar spacers of several lengths—162 mm, 110 mm, and 44 mm—which allowed correction for dispersion of phase change. Plate finesses of 40 were typical. The parallelism of the plates was adjustable from outside the Fabry-Perot vacuum chamber. In order to randomize errors resulting from nonflatness and misalignment of the plates, the parallelism of the plates was adjusted between each scan. The fringes were scanned in a nearly linear fashion by leaking dry nitrogen into the Fabry-Perot chamber through a supersonic nozzle. By recording several fringes, small corrections for residual nonlinearity of the scan rate were evaluated.

Signals from the photomultipliers were displayed on a strip chart recorder and were also digitized and stored in a small computer. A typical strip chart recording is shown in Fig. 14. At the end of a scan the data were transferred from the computer to punched paper tape. A scan of three laser fringes and three krypton fringes took approximately 20 min to record and consisted of approximately 1000 points on each line.

A general purpose computer was used to fit the digitally recorded signals and to determine fractional orders of interference. The laser fringes were sufficiently narrow (~5 MHz wide) that a recording of their profile very nearly approximated the instrumental profile. For these fringes the functional form of the profile was assumed to be a Lorentzian convoluted with an aperture function. The width of the aperture function was calculated using the focal length of the collimating lens L_2 and the diameter of aperture A_2 (0.287-mm diam) and was not varied by the computer program. The computer program fit each laser order with the profile having an adjustable position, intensity, and Lorentzian width. In addition, an adjustable background intensity for the entire scan was included. For a typical scan of three laser fringes a total of ten adjustable parameters were varied by the computer in an iterative fashion to obtain a best fit in the least-squares sense. Figure 15 illustrates the quality of fit for a typical laser fringe.

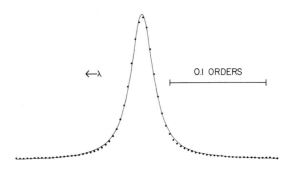

Fig. 15. Computed fit of one order of the ^{3}He-^{20}Ne:^{129}I$_2$ k line at 633 mm. The spacer length was 44 mm. The dots represent the data points, and the solid line is the computed spectrogram.

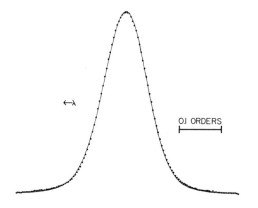

Fig. 16. Computed fit of one order of the ^{86}Kr 606-nm line. The spacer length was 44 mm. The dots represent the data points, and the solid line is the computed spectrogram.

For the krypton fringes the functional form of the profile was assumed to be the sum of two Gaussians convoluted with the instrumental profile (a Lorentzian convoluted with the aperture function) that was determined from the laser fringes. The sum of two Gaussians was used to take into account the slight asymmetry of the krypton profile. The spacing and relative intensity of the two krypton components were assumed to be those determined by Rowley and Hamon,[22] the weaker component being 0.007 cm^{-1} to the red of and 0.06 times as intense as the stronger component. The fitting for a krypton scan was identical to the fitting of a laser scan except each krypton fringe was fit with the krypton profile having an adjustable position, intensity, and Gaussian width. (The widths of the stronger and weaker components were required to be equal.) The computer fitting of a typical krypton fringe is shown in Fig. 16.

The absolute wavelength of the ^{129}I$_2$ k peak was determined from twenty-three pairs of fractional orders, a pair of fractional orders being one fractional order each of ^{129}I$_2$ and ^{86}Kr recorded simultaneously. All fractional orders were measured relative to a reference pressure of 0.025 Torr. The number of measured pairs of fractional orders for the 162-mm, 110-mm, and 44-mm spacers is seven, eight and eight, respectively. The ^{86}Kr 606-nm primary standard was the only standard wavelength used in this measurement, but one scan of the ^{86}Kr 565-nm line was recorded for each spacer so that ambiguity in the determination of integral order numbers could be eliminated. The twenty-three pairs of order numbers are recorded in Table I and were computer fit with a straight line as explained in detail in Ref. 23. From the slope of this line, which was determined with a statistical uncertainty of 1.4 parts in 10^9, the wavelength of the ^{129}I$_2$ k peak was determined.

The slight asymmetry of the ^{86}Kr 606-nm primary standard causes ambiguity in the assignment of the defined wavelength, 605 780.2105 pm, to the krypton profile. If the defined wavelength is assigned to the maximum of the krypton profile, the wavelength of the ^{129}I$_2$ k peak is 632 991.2758 ± 0.0009 pm. If the defined wavelength is assigned to the center of gravity of the krypton profile, the wavelength of the ^{129}I$_2$ k peak is 632 991.2670 ± 0.0009 pm. The uncertainties are statistical errors only and represent one standard deviation.

The recommendation of the CIPM states that the wavelengths of Kr lamps made in conformity with its recipe[19] will agree within 1 part in 10^8 of the wavelength corresponding to the transition between unperturbed levels in ^{86}Kr. This recommendation was adopted before evidence of an asymmetric profile for this line was available. Since that time, experience appears to indicate that a large part of the uncertainty in realizing the standard wavelength is removed by assuming that the krypton profile has an asymmetry similar to that proposed by Rowley and Hamon. Some of the recent measurements that provide information on the reproducibility of krypton lamps include (a) the comparison of two krypton lamps from different laboratories by Barger and Hall[24] in which a wavelength difference of 1 ± 2.5 parts in 10^9 was found, (b) the comparison of four krypton lamps at BIPM,[20] which all produced the same wavelength to within 1 part in 10^9, (c) the close agreement (2.8 parts in 10^9) of wavelength measurements of the ^{127}I$_2$ i peak at the National Research Council in Canada,[5] at BIPM,[20] at NPL,[25,26] and at NBS as described in a later section of this paper, and (d) the comparison of the methane wavelength as measured by direct comparison with krypton at NBS (Boulder)[24] and at BIPM,[20] as measured at NRC using frequency mixing techniques,[27] and as

Table I. Measured Order Numbers

Date	Spacer (mm)	Standard order number[a]	Iodine order number
1	162	536 243.42165	513 191.36334
1	162	536 243.32643	513 191.27089
2	162	536 243.04272	513 191.00252
2	162	536 243.50969	513 191.44803
2	162	536 243.51521	513 191.45414
3	162	536 243.40107	513 191.34347
3	162	536 243.41134	513 191.35585
4	110	364 572.41939	348 900.16590
4	110	364 572.40389	348 900.15140
5	110	364 572.15685	348 899.91757
5	110	364 572.27147	348 900.02602
5	110	364 572.31335	348 900.06562
6	110	364 572.22004	348 899.97403
6	110	364 572.36289	348 900.11231
6	110	364 572.41445	348 900.16217
7	44	146 881.52530	140 567.38345
8	44	146 881.41806	140 567.28067
8	44	146 881.46117	140 567.32146
8	44	146 881.53128	140 567.38863
8	44	146 881.56608	140 567.42179
8	44	146 881.57849	140 567.43434
9	44	146 881.49394	140 567.35242
9	44	146 881.44020	140 567.30081

[a] The order numbers given apply to the larger component of the krypton profile.

measured by combining the $^{129}I_2$ k peak wavelength reported here with the iodine-methane wavelength ratio that is currently being measured. The three completed comparisons agree to within 3 parts in 10^9, and preliminary results from the ratio experiment indicate that all four methane wavelengths will agree within ~ 3 parts in 10^9.

There are two sources of systematic uncertainty associated with the krypton lamp. One deals with the choice of a particular model for the asymmetry, while the other deals with the imperfect realization of the krypton wavelength that defines the meter. Because the magnitude of the wavelength uncertainties associated with the choice of model and the imperfect realization of the Kr wavelength are difficult to estimate, we choose to base our wavelength measurement on a particular model, that of Rowley and Hamon, and on a particular realization of the krypton standard. Thus we have not included any systematic error for the model or the imperfect realization of the Kr standard.

We have attempted to check the validity of the Rowley and Hamon model by fitting some of the krypton profiles at each spacer length with slightly different two component models. Because the asymmetry is very slight and the recordings of the krypton fringes are somewhat noisy, this method does not provide a distinctly best two-component model. Some of the effects of the noise could be alleviated by averaging and smoothing the recorded krypton profiles. However, there is a limit to the usefulness of this technique without including parameters in the instrument function that take into account small defects in the interferometer plates. The contribution to the asymmetry from such effects would probably be smaller than that of the spectral profile but large enough to confuse the choice of a best profile.

^{3}He–^{22}Ne:$^{129}I_2$ B

The absolute wavelength of He–Ne lasers stabilized to other iodine peaks can be determined by measuring the difference frequency between them and a laser stabilized to the $^{129}I_2$ k peak and combining this difference frequency with the wavelength of the k peak reported above. The ^{3}He–^{22}Ne laser line is coincident with a conspicuous $^{129}I_2$ hyperfine triplet. We have stabilized the ^{3}He–^{22}Ne laser on the center component (hereafter called the B component) of the triplet and measured the beat frequency between it and the $^{129}I_2$ k peak.

Both lasers were locked using first derivative signals, and the slopes of the backgrounds were adjusted to be zero. Light from the two lasers was incident on a photomultiplier, and the beat note was detected by an appropriately tuned microwave spectrum analyzer. The tenth harmonic frequency from a variable frequency generator was made to coincide with the beat note and the divide by 10 output of the frequency generator was counted. Direct counting of the beat frequency was not possible because the beat frequency was higher than what our counter would accept. The two lasers were synchronously modulated to decrease the width of the beat note to approximately 1 MHz.

Ten measurements of the beat note frequency were made. Between successive measurements both lasers were unlocked and the background slopes adjusted to zero. The average beat note frequency is 892.467 ± 0.083 MHz with the k peak being the lower frequency. Combining this result with the $^{129}I_2$ k peak wavelength for the center of gravity definition of the krypton profile gives a wavelength for the ^{3}He–^{22}Ne laser stabilized to the $^{129}I_2$ B peak of 632 990.0742 ± 0.0009 pm.

^{3}He–^{20}Ne:$^{127}I_2$ i

The most interesting iodine wavelength that can be determined by beating against a laser stabilized to the $^{129}I_2$ k peak is that of the ^{3}He–^{20}Ne laser stabilized to the $^{127}I_2$ i peak. The absolute wavelength of the $^{127}I_2$ i peak has been measured interferometrically at the National Research Council of Canada,[5] at BIPM,[20] and at NPL.[25,26] Measurement of this beat note provides a desirable coupling beween wavelength measurements being made at four national standards laboratories. Because the background under the $^{127}I_2$ i peak cannot be made zero, third derivative locking techniques were essential for the $^{127}I_2$ laser. The $^{129}I_2$ laser was also locked using third derivative signals, but this was a matter of convenience and was not essential. For accuracies on the order of 5 in 10^{10}, which are all that are needed for this comparison, third derivative locking on a background having a nonzero slope and first derivative locking on a background having a zero slope are equivalent.

Light from the two lasers was incident on the photomultiplier, and the beat frequency was amplified and counted. The lasers were synchronously modulated, and the beat note was approximately 1 MHz wide. The average beat note frequency was 96.0399 ± 0.0098 MHz with the k peak having the higher frequency. The wavelength for the ^{3}He–^{20}Ne laser stabilized to the $^{127}I_2$ i peak is thus found to be 632 991.3954 ± 0.0009 pm, where the defined krypton wavelength value is assigned to the center of gravity of the krypton profile.

The BIPM value for the $^{127}I_2$ i peak is 632 991.3960 ± 0.0012 pm and can be compared directly to the value reported above because the defined value for the krypton standard has also been associated with the center of gravity of the krypton profile.

The NRC and NPL values for the wavelength of the $^{127}I_2$ i peak are 632991.398 ± 0.003 pm and 632 991.3990 ± 0.0008 pm, respectively. The NRC value is based on the assumption that the defined value for the krypton standard is to be associated with a point intermediate between the maximum intensity point and the center of gravity of the krypton profile.[28] In the NPL measurement no correction has been made for asymmetry, so it is impossible to determine the point on the krypton profile that has been assigned the defined krypton wavelength. However, since it is probably a point somewhere between the maximum intensity and center of gravity, the NPL assignment of the krypton wavelength to the krypton profile is probably similar to that used by NRC. If we use the NRC assumption for the defined value for the krypton standard, the NBS value for the $^{127}I_2$ i peak is found to be 632 991.3998 ± 0.0009 pm, which agrees satisfactorily with the NRC and NPL values.

Other Wavelengths

In the course of these measurements we obtained the frequency intervals between the above components and several other components in the $^{129}I_2$ systems. The intervals measured were the A-B, B-C, and k-n intervals. From these intervals the following absolute wavelengths were deduced: $^{129}I_2$ component A = 632 990.1007 ± 0.0009 pm, $^{129}I_2$ component C = 632 990.0502 ± 0.0009 pm, $^{129}I_2$ component n = 632 991.3414 ± 0.0009 pm. As before, the defined value for the krypton standard has been assigned to the center of gravity of the krypton profile.

CCDM Recommended Wavelengths

At the June 1973 meeting, the Comité Consultatif pour la Définition du Mètre (CCDM) recommended wavelength values for He–Ne lasers stabilized to components of methane and iodine lines.[29] These values are based on the assumption that the defined value for the krypton wavelength is to be assigned to a point on the krypton profile midway between the peak and center of gravity.[30] In terms of this midpoint definition our measured wavelengths become
^{3}He–^{20}Ne:$^{129}I_2$, k λ = 632 991.2714 ± 0.0009 pm,
^{3}He–^{22}Ne:$^{129}I_2$, B λ = 632 990.0786 ± 0.0009 pm,
and ^{3}He–^{20}Ne:$^{127}I_2$, i λ = 632 991.3998 ± 0.0009 pm.
Several laboratories have made measurements on the $^{127}I_2$, i peak, and a rounded wavelength value for this peak has been recommended by CCDM. The recommended value is λ = 632 991.399 ± 0.0025 pm.

Comparison with Wavelength Measurements on the 3.39-μm Line

The only other laser line whose wavelength has been measured with a precision comparable to the iodine stabilized laser lines is the 3.39-μm line of the He–Ne laser stabilized to methane. Wavelength values for this line have recently been reported by Barger and Hall[24] and Giacomo.[20] A frequency controlled Fabry-Perot was used by the former, while a Michelson interferometer was used by the latter. In both experiments, the 3.39-μm line was compared directly with the krypton primary standard. Accuracies of 3.5 and 5.0 parts in 10^9 have been achieved by the former and latter groups, respectively. The frequency mixing techniques of Baird *et al.*[27] provide another independent determination of the methane wavelength. By combining these wavelength values with the frequency measurement of the 3.39-μm line by Evenson *et al*[31] precise values for the speed of light have been obtained. Hall and Bordé[32] have recently reported observation of magnetic hyperfine structure in this methane line, which can lead to small, intensity dependent, frequency shifts. Such shifts are too small to be observable in any of the wavelength comparisons considered here.

The close proximity of the iodine and krypton lines as compared with the methane and krypton lines, makes the iodine wavelength measurement easier in two respects. First, the narrow iodine laser line provides an accurate recording of the instrument window, which is a significant advantage in treating the asymmetric krypton profile. Second, higher finesse interferometers are available ($\sim$40 vs $\sim$10), which reduces instrumental broadening.

Summary

We have given a description of lasers stabilized to components of the $^{129}I_2$ spectrum in the region of the 633-nm laser lines for ^{3}He–^{20}Ne and ^{3}He–^{22}Ne. We have shown how the operational characteristics such as power output, peak size, and width, are related to each other and to some of the controllable parameters such as excitation level, iodine absorption, and iodine pressure. These lasers exhibit a frequency stability for 10-sec sampling time of about 2×10^{-12} ν and a resetability of about 1×10^{-10} ν.

The absolute vacuum wavelengths for several iodine components have been measured against the ^{87}Kr standard. Among these are
^{3}He–^{20}Ne:$^{129}I_2$, k λ = 632 991.2670 ± 0.0009 pm,
^{3}He–^{22}Ne:$^{129}I_2$, B λ = 632 990.0742 ± 0.0009 pm,
^{3}He–^{20}Ne:$^{127}I_2$, i λ = 632 991.3954 ± 0.0009 pm.
These results were obtained on the assumption that the defined value for the krypton standard was to be associated with the center of gravity of the line profile.

References

1. P. H. Lee and M. L. Skolnick, Appl. Phys. Lett. **10**, 303 (1967).
2. R. L. Barger and J. L. Hall, Phys. Rev. Lett. **22**, 408 (1969).
3. G. R. Hanes and C. E. Dahlstrom, Appl. Phys. Lett. **14**, 362 (1969).
4. A. J. Wallard, J. Phys. **E5**, 926 (1972).
5. G. R. Hanes, K. M. Baird, and J. DeRemigis, Appl. Opt. **12**, 1600 (1973).
6. J. D. Knox and Y. H. Pao, Appl. Phys. Lett. **18**, 360 (1971).
7. H. P. Layer, R. D. Deslattes, and W. G. Schweitzer, Jr. (to be published).
8. Our $^{129}I_2$ was obtained from the Nuclear Division, Union Carbide Company, Oak Ridge National Laboratory. The isotopic purity is said to be about 75% $^{129}I_2$, 25% $^{127}I_2$.
9. Polyethylene terephthalate.
10. 1 Torr $\approx$ 133 N/m².
11. We have learned from Y. H. Pao that he has nearly completed an effort at classification of this group of components. His iodine sample contained about 20% iodine 127 and 80% iodine 129. According to Pao the observed spectrum is an overlay of components due to $^{129}I_2$, $^{127}I_2$, and $^{127,129}I_2$.
12. G. R. Hanes, J. LaPierre, P. R. Bunker, and K. C. Shotton, J. Mol. Spectrosc. **39**, 506 (1971).
13. J. D. Knox and Y. H. Pao, Appl. Phys. Lett. **16**, 129 (1970).
14. For a discussion of hysteresis in lasers with internal absorption cells see H. Greenstein, J. Appl. Phys. **43**, 1732 (1972). This paper also treats many other aspects of such lasers.

15. We have learned that J. B. Cole of the National Standards Laboratory, Australia, is approaching the problem of limited operating range in his iodine stabilized laser by using a three-mirror laser cavity as a means of varying the interaction of the absorber with the oscillating mode. (Submission to 5e session, Comité Consultatif pour la Definition du Mètre.)
16. P. W. Smith, J. Appl. Phys. **37,** 2089 (1966).
17. J. A. Barnes, A. R. Chi, L. S. Cutler, D. J. Healy, D. B. Leason, T. E. McGunigal, J. A. Mullen, Jr., W. L. Smith, R. L. Sydnor, R. F. C. Vessot, and G. M. R. Winkler, IEEE trans. Instrum. Meas. **IM-20,** 105 (1971).
18. Comité Intern. des Poids et Measures, Procès-Verbaux des Séances 2e seri **28,** 70, 1960.
19. Reference 18, p. 71.
20. Doc. CCDM/73-15, BIPM, 5e session, 1973.
21. K. M. Baird and D. S. Smith, J. Opt. Soc. Am. **52,** 507 (1962).
22. W. R. C. Rowley and J. Hamon, Rev. Opt. **42,** 519 (1963). Rowley has further discussed the symmetry of the krypton line in Doc. CCDM/73-28, NPL, 5^e session, 1973.
23. E. G. Kessler, Jr., Phys. Rev. **A7,** 408 (1973).
24. R. L. Barger and J. L. Hall, Appl. Phys. Lett. **22,** 196 (1973).
25. Doc. CCDM/73-14, NPL, 5^e session, 1973.
26. W. R. C. Rowley and A. J. Wallard, J. Phys. **E6,** 647 (1973).
27. K. M. Baird, D. S. Smith, and W. E. Berger, Opt. Commun. **7,** 107 (1973).
28. Doc. CCDM/73-11, NRC, 5e session, 1973.
29. CCDM Recommendation Ml (1973).
30. K. G. Kessler, NBS; private communication.
31. K. M. Evenson, J. S. Wells, F. R. Peterson, B. L. Danielson, and G. W. Day, Appl. Phys. Lett. **22,** 192 (1973).
32. J. L. Hall and C. Bordé, Phys. Rev. Lett. **30,** 1101 (1973).

Determination of the speed of light by absolute wavelength measurement of the R(14) line of the CO_2 9.4-μm band and the known frequency of this line

J.-P. Monchalin, M. J. Kelly, J. E. Thomas, N. A. Kurnit, A. Szöke, and A. Javan

Department of Physics, Massachusetts Institute of Technology, Cambridge, Massachusetts 02139

F. Zernike

Perkin-Elmer Corporation, Norwalk, Connecticut

P. H. Lee

Physics Department, University of California, Santa Barbara, California

Received April 12, 1977

A precision long-arm scanning Michelson interferometer system is described that is capable of measuring absolute laser wavelength to within several parts in 10^9 in the 10-μm spectral range and to within several parts in 10^{11} in the visible range. The R(14) line of the CO_2 9.4-μm band is measured to be 9.305 385 613 (70) μm. This measured value and the known frequency of this line give a value for the speed of light: c = 299 792 457.6 (2.2) m/sec, in agreement with the recent independent measurements of c and its recommended value.

Since the late 1960's, in a continuing experimental program at MIT, a precision scanning long-arm vacuum Michelson interferometer has been developed and perfected for an accurate comparison of two widely different laser wavelengths, one of them lying in the far infrared or the infrared and the other in the visible region. In the experiment, the visible laser is a frequency-stabilized He–Ne 633-nm laser having an accurately calibrated wavelength with respect to a Kr standard. The measurements are performed by simultaneous fringe counting and relative fringe-phase comparison at the two wavelengths, while the scanning arm of the interferometer is varied over a path length of about 50 cm. The precise absolute wavelength of the far-infrared or the infrared laser is obtained from this simultaneous fringe counting and the calibrated wavelength of the He–Ne laser. On-line data processing has made possible measurements of relative phases to within a small fraction of the He–Ne red fringe. The limiting accuracy of these measurements is set by the ability to make correction for the systematic fringe shift caused by diffraction. These shifts were minimized by using large-aperture optical components. Other major practical limitations to the accuracy arise from the laser-beam quality, the quality of optical surfaces, and the ability to align the two laser beams collinearly.

Over the past several years, the design of the interferometer and the measuring procedures have been refined to obtain higher accuracy.[1,2] Our accuracy limit at this time is several parts in 10^9 in the 10-μm region of the spectrum. In the visible region (where the diffraction fringe shift is appreciably less), the interferometer can be adapted and applied to the absolute laser-wavelength measurements to within a few parts in 10^{11}. Because of its low-Q and broadband operating capability, it can be used as a broadband spectrometer for precise spectroscopic studies through accurate laser wavelength measurements over the entire wavelength region permitted by its transmitting optics (the beamsplitter, the windows, and the compensator). An application of this has recently been reported[3] in a spectroscopic study of the CO_2 01^11–$(11^10,03^10)_I$ band, which oscillates in the 11.2-μm region.

The first use of this interferometer in a precision determination of c by simultaneous measurements of the absolute laser wavelength and frequency was reported[1] in 1969. In that experiment the interferometer was used to measure the absolute wavelength of a D_2O 84-μm laser radiation: this, together with the precise measured frequency[4] of that laser line, gave a value for the speed of light to within an accuracy of 2 parts in 10^6, comparable at that time with the best previous measurements. This Letter reports the application of the interferometer with the improved precision to an independent determination of c to within an accuracy of 7 parts in 10^9; the measurement is done by precise wavelength determination of the center of the Doppler-free resonance of the R(14) line of the CO_2 9.4-μm $(00°1)$–$(10°0, 02°0)_{II}$ band.[5] The known absolute frequency of this line[6] and the measured wavelength give a precise value of c. Our measured value agrees with the recent measurements of c with a comparable accuracy.[7]

Figure 1 is a block diagram of the experiment. An important feature of the interferometer is the use of a flat mirror on one arm and a corner reflector on the variable arm: with this configuration the two beams having different wavelengths can be made accurately parallel, since flat interference patterns are obtained only when the beams corresponding to each wavelength are perpendicular to the flat mirror.

The wavelength measurements reported here were performed using a frequency-stabilized CO_2 laser oscillating at the center of the narrow Doppler-free resonance belonging to the R(14) line. This Doppler-free resonance is obtained by a well-known method[8] in which

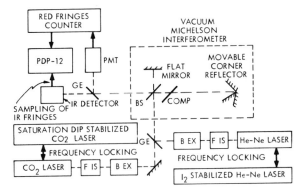

Fig. 1 Simplified block diagram of the experimental setup: BS = beam splitter, COMP = compensator, GE = germanium slab (reflects the red and transmits the infrared), BEX = beam expander, FIS = feedback-isolation optics (polarizer, quarter-wave plate, attenuator). All the optics have a large size with a useful aperture of 50 mm. The corner reflector carriage rides on two carefully polished stainless steel rods, and its surface in contact with the rods is coated with Teflon. The translation is obtained by the motion of an ac-synchronous motor with this motion transferred to the carriage by the use of gears, pulleys, and a strong steel cable. A heavy flywheel (inside the vacuum-interferometer assembly) attached to the main pulley gives a highly uniform and jitter-free translation over a length of about 50 cm.

the laser-induced fluorescence at the $(001) \rightarrow (000)$ 4.3-μm CO_2 emission band is used in the detection system. Resonance full widths of 1 part in 10^8 were typical in this experiment.[9] The laser frequency was stabilized at the line center to within one tenth of the resonance width.

In order to avoid fringe modulation, the CO_2 laser used in the measurement was stabilized to its zero beat with respect to a frequency-modulated laser, which was first-derivative locked to the center of the $R(14)$ Doppler-free resonance.

The He-Ne laser used was likewise stabilized to the center frequency of a second laser locked to the inverted Lamb dip obtained with a low-pressure intracavity iodine cell. Two He-Ne lasers similarly and independently locked to the center frequency of the iodine inverted Lamb dip were used to determine the He-Ne laser frequency resetability. From the beat note obtained by mixing the output of the two lasers, their frequency resetability was estimated[10] to be ±5 parts in 10^{10}.

The largest systematic error requiring correction arises from the fringe shift due to diffraction of the spatially limited beam propagating in the interferometer (without an aperture in its beam path). Since this error scales as λ^2, the dominant diffraction correction is introduced by the CO_2 infrared beam.

It is known that, for the lowest-order propagation mode consisting of a TEM_{00} Gaussian beam, this correction is given by $(\lambda - \lambda_{\exp})/\lambda = -\lambda^2/(4\pi^2 w_0^2)$, where $\lambda_{\exp}$ is the experimental value given by the ratio of the fringe counts and w_0 is the beam radius at $1/e$ of its E-field distribution.[11,12]

With the laser optimally aligned for operation on its lowest-order mode, the beam profile was carefully measured at the input of the interferometer. It was found that this beam profile reproducibly and dominantly consisted of the TEM_{00} mode with a small admixture of the high-order modes in such a way as to cause a slightly astigmatic beam with, in fact, perfect collimation along its two principal axes. (This profile was characteristic of the CO_2 laser used in the experiment; it utilized one Brewster-angle polarizing plate in its cavity.) It can be shown that, for this astigmatic beam, the diffraction correction has the same form as the TEM_{00} mode given above after substituting for $1/w_0^2$, the quantity $\frac{1}{2}(1/w_{0x}^2 + 1/w_{0y}^2)$, where w_{0w} and w_{0y} are the beam radii along the two principal directions.[13]

Since the corner reflector is used at its center, the small empty spaces existing between its adjacent mirrors cause diffraction resulting in an additional small correction. This correction can be estimated by the scalar diffraction theory.[13] Correcting for this and the diffraction effect, one obtains

$$\frac{\lambda - \lambda_{\exp}}{\lambda} = -\frac{\lambda^2}{8\pi^2}\left(\frac{1}{w_{0x}^2} + \frac{1}{w_{0y}^2}\right)\left(1 - \frac{9a_0}{2\sqrt{\pi}w_0}\right),$$

wherein the last factor, a_0, is the width of the empty space at the junction of the corner reflector's mirrors, which is much smaller than the quantity w_0 given by $(w_{0w} + w_{0y})/2$. Inspection shows that other diffraction fringe shifts are much smaller, and their contributions lie below our experimental error.

The measurements of the beam widths were performed with a detector, a pinhole, and a two-dimensional beam-steering mechanism displaying the CO_2 beam profile at the input of the interferometer. The spacing a_0 was estimated from a large-scale photograph of the corner reflector.

From the above studies, the final diffraction correction is found to be $[(\lambda - \lambda_{\exp})/\lambda] = -1.89 \times 10^{-8}$, corresponding to $(\lambda - \lambda_{\exp}) = -1.76 \times 10^{-7}$ μm. The uncertainty of this result is conservatively estimated to be ±20% according to the following: 8% comes from w_{0x} and w_{0y} measurements (which were performed with ±4% precision); another 8% originates from a_0 determination (which was measured with ±25% uncertainty for $w_0 \approx 10$ mm and $a_0 \approx 0.6$ mm).

The experimental results are plotted in Fig. 2. For each point of the plot, the red and infrared fringes have been maximized before scanning. For each group of six data points (separated by vertical lines on the plot), the red and infrared beams have been carefully recentered on the corner reflector for any position along the scan by realigning the entire interferometer. This centering on the corner reflector (for any position along the scan) was used to obtain the required parallelism of the red and the infrared beams with respect to the axis of the reflector translation. In the absence of this parallelism, as the interferometer arm is varied, the wavefront of the beam reflected by the corner reflector moves across the fixed wavefront of the beam reflected by the flat mirror; this introduces an error if these wavefronts are not perfectly flat. The experimental scatter seen in Fig. 2 is in part caused by this effect. (The slight deviation from a perfectly flat wavefront is caused by the lack of

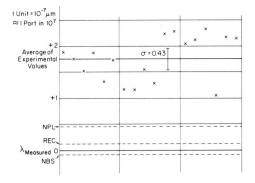

Fig. 2 Plot of the experimental results obtained directly from the ratio of the fringe counts. The origin of the vertical axis is chosen at the final result ($\lambda_{\text{MEASURED}}$), which is obtained by subtracting the diffraction correction (1.76×10^{-7} μm) from the average of the experimental data. One standard deviation ($\sigma = 0.43\ 10^{-7}$ μm) is indicated by an error bar. The lines marked NPL, NBS, and REC correspond, respectively, to values of the wavelength of the $R(14)$ line calculated by using the values of the speed of light given by the National Physical Laboratory, Teddington (England), the National Bureau of Standards, Boulder (Colorado) and the recommended value of c (see Ref. 7).

perfect flatness of our beam splitter and the compensator, which were made of sodium chloride.) The precision of each data point is typically 1–2 parts in 10^9, as determined by a least-squares fit to a straight line of the relative fringe phases determined at 38 equally spaced intervals during a scan.

In the measured value $\lambda_{\text{MEASURED}}$, we use the recommended[14] value for the i component of iodine-127: 0.632 991 399 μm (uncertainty ± 4 parts in 10^9). The experimental standard deviation is 4.3×10^{-8} μm. By combining quadratically the standard deviation and the uncertainty of the diffraction correction, one finds that the uncertainty of the ratio $\lambda_{\text{CO}_2}/\lambda_{\text{He-Ne}}$ is ± 6 parts in 10^9. The uncertainty of $\lambda_{R(14)}$ is $\pm 7.2\ 10^{-9}$, and the result is found to be: $\lambda_{R(14)} = 9.305\ 385\ 613\ (70)$ μm.

This gives a value for the speed of light $c = 299\ 792\ 457.6\ (2.2)$ m/s (relative uncertainty ± 7.3 parts in 10^9). This is in excellent agreement with the recommended[14] value of c (299 729 458 m/sec) based on the previous independent measurements of this quantity.[7]

The ratio found for $\lambda_{R(14)}/\lambda_{I_2(i)}$ also compares very well with the ratio of their frequencies calculated from the frequency of the $R(14)$ line and the frequency of the i component of the iodine transition deduced from the recent wavelength comparison[15] of an iodine-stabilized He–Ne 633-nm laser with the wavelength of a methane-stabilized 3.39-μm He–Ne laser. The difference between the two ratios is 2.8 parts in 10^9.

This work was supported by the Air Force Cambridge Research Laboratories, the National Science Foundation, and the U.S. Army Research Office, Durham, North Carolina. J.-P. Monchalin is now at Ecole Polytechnique, Université de Montréal; M. J. Kelly at Physics Department, University of California, Berkeley; N. A. Kurnit at Los Alamos Scientific Laboratory, Los Alamos, New Mexico; A. Szöke at Lawrence Livermore Laboratory, University of California, Livermore; and F. Zernike at Philips Laboratories, Briarcliff Manor, New York. J. E. Thomas is a Hertz Predoctoral Fellow.

References

1. V. Daneu, L. O. Hocker, A. Javan, D. Ramachandra Rao, A. Szöke, and F. Zernike, Phys. Lett. **29A**, 319 (1969).
2. J.-P. Monchalin, A. Javan, N. A. Kurnit, A. Szöke, and F. Zernike, Bull. Am. Phys. Soc. **16**, 1403 (1971); N. A. Kurnit, in *Fundamental and Applied Laser Physics*, Proceedings of the Esfahan Symposium, 1971, M. S. Feld, A. Javan, and N. A. Kurnit, eds. (Wiley-Interscience, New York, 1973), p. 479; J.-P. Monchalin, M. J. Kelly, J. G. Small, F. Keilmann, N. A. Kurnit, A. Javan, and F. Zernike, *Proceedings of the Conference on Precision Electromagnetic Measurements* (Boulder, Colorado, 1972); J.-P. Monchalin, M. J. Kelly, J. E. Thomas, N. A. Kurnit, and A. Javan, *Proceedings of the Atomic Spectroscopy Symposium 1975* (NBS, Washington, D.C.).
3. J.-P. Monchalin, M. J. Kelly, J. E. Thomas, N. A. Kurnit, and A. Javan, J. Mol. Spectrosc. **64**, 491 (1977).
4. L. O. Hocker, J. G. Small, and A. Javan, Phys. Lett. **29A**, 321 (1969).
5. The reason for the choice of this $R(14)$ line relates to the frequency-measurements chain. See A. Javan in *Fundamental and Applied Laser Physics*, p. 295.
6. K. M. Evenson, J. S. Wells, F. R. Petersen, B. L. Danielson, and G. W. Day, Appl. Phys. Lett. **22**, 192 (1973); T. G. Blaney, C. C. Bradley, G. J. Edwards, D. J. E. Knight, P. T. Woods, and B. W. Jollife, Nature **244**, 504 (1973); F. R. Petersen, D. G. McDonald, J. D. Cupp, and B. L. Danielson, *Laser Spectroscopy*, Proceedings of the Vail (Colorado) Conference, R. G. Brewer and A. Mooradian, eds., p. 555.
7. K. M. Evenson, J. S. Wells, F. R. Petersen, B. L. Danielson, G. W. Day, R. L. Barger, and J. L. Hall, Phys. Rev. Lett. **29**, 1346 (1972); T. G. Blaney, C. C. Bradley, C. J. Edwards, B. W. Jollife, D. J. E. Knight, W. R. C. Rowley, K. C. Shotton, and P. T. Woods, Nature **251**, 46 (1974).
8. C. Freed and A. Javan, Appl. Phys. Lett. **17**, 53 (1970). For an experimental arrangement similar to the one used here see M. J. Kelly, J. E. Thomas, J.-P. Monchalin, N. A. Kurnit, and A. Javan, "Observation of anomalous Zeeman effect in infrared transitions of $^1\Sigma$ CO_2 and N_2O molecules," Phys. Rev. Lett. **37**, 686 (1976).
9. Much narrower linewidths have been observed when the laser beam is expanded and the pressure reduced. See M. J. Kelly, J. E. Thomas, J.-P. Monchalin, N. A. Kurnit, and A. Javan, *Proceedings of the 29th Annual Symposium on Frequency Control* (1975), and M. J. Kelly, Ph.D. Thesis (MIT, 1976) (unpublished).
10. A first-derivative locking technique was used. An improvement by more than one order of magnitude can readily be obtained.
11. H. Kogelnik and T. Li, Appl. Opt. **5**, 1550 (1966).
12. J.-P. Monchalin, M. J. Kelly, J. E. Thomas, N. A. Kurnit, A. Szöke, F. Zernike, P. H. Lee, and A. Javan, *Frontiers in Laser Spectroscopy*, Les Houches, 1975, Session 27 (North-Holland, Amsterdam, 1976).
13. J.-P. Monchalin, Ph.D. Thesis (MIT, 1976) (unpublished).
14. Fifth Meeting of the Comité Consultatif pour la Définition du Mètre, Metrologia **10**, 75 (1974); J. Terrien, Nouv. Rev. Opt. **4**, 215 (1973).
15. H. P. Layer, R. D. Deslattes, and W. G. Schweitzer, Jr., Appl. Opt. **15**, 734 (1976).

Determination of the speed of light by absolute wavelength measurement of the R(14) line of the CO_2 9.4-μm band and the known frequency of this line: errata

J.-P. Monchalin, M. J. Kelly, J. E. Thomas, N. A. Kurnit, A. Szöke, and A. Javan

Department of Physics, Massachusetts Institute of Technology, Cambridge, Massachusetts 02139

F. Zernike

Perkin-Elmer Corporation, Norwalk, Connecticut

P. H. Lee

Physics Department, University of California, Santa Barbara, California

Received August 23, 1977

In our recent Letter,[1] the label of the vertical axis of the plot of Fig. 2 should read:

$$1 \text{ Unit} = 10^{-7} \mu\text{m} \approx 1 \text{ Part in } 10^8.$$

The recommended value of the speed of light, c, should read:

$$299\ 792\ 458 \text{ m/sec}.$$

Reference

1. J.-P. Monchalin, M. J. Kelly, J. E. Thomas, N. A. Kurnit, A. Szöke, A. Javan, F. Zernike, and P. H. Lee, Opt. Lett. 1, 5 (1977).

Proc. R. Soc. Lond. A. 355, 61–88 (1977)
Printed in Great Britain

Measurement of the speed of light
I. Introduction and frequency measurement of a carbon dioxide laser

By T. G. Blaney, C. C. Bradley, G. J. Edwards, B. W. Jolliffe, D. J. E. Knight, W. R. C. Rowley, K. C. Shotton and P. T. Woods

National Physical Laboratory, Teddington, Middlesex, U.K.

(*Communicated by J. Dyson, F.R.S. – Received 6 October* 1976)

[Plates 1 and 2]

This, and part II following, describe a determination of the speed of light made by measuring the frequency and wavelength of radiation from a CO_2 laser. This laser was operated on the 9.3 μm R(12) transition, and stabilized by reference to fluorescence in an external CO_2 absorption cell. The laser frequency was measured via a sequence of harmonic-mixing stages involving the HCN and H_2O lasers as transfer oscillators, and was found to be 32 176 079 482 ± 14 kHz (± 4.2 parts in 10^{10}). A correction to this value of −10 kHz is necessary to obtain the centre frequency of the CO_2 reference transition. The wavelength measurements and value of c are described in part II.

1. Introduction – the speed of light

The speed of light is one of the most important of the fundamental constants, having particular significance in electromagnetic, atomic and relativity theory as well as practical application in the determination of astronomical and terrestrial distances. There has been much interest in its measurement and a progressive demand for more accurate values.

The earliest estimate of the speed of light was made by Roemer in 1676 from changes of the time of eclipse of a moon of Jupiter as the distance of the Earth from that planet varied. Since then, and particularly since the end of the nineteenth century, many measurements have been made. These are reviewed comprehensively in a book by Froome & Essen (1969). In addition to direct measurements with visible light, measurements have also been made on radio waves and by indirect methods such as that involving the ratio of electromagnetic to electrostatic units. Experiments of particular significance, due to the accuracies achieved, were those of Essen (1950) using a cavity-resonator, Bergstrand (1951) using modulated light, and Froome (1958) using a 4 mm microwave interferometer. These led to the adoption in 1957 of a value of 299 792.5 km/s as the conventional value for the velocity of plane electromagnetic waves in free space, c. During the last few years a further series of measurements of significantly greater accuracy has been made by using lasers and, as discussed in the conclusion to part II of this paper, a value of 299 792 458 m/s is at present recommended for general use.

There were several reasons why this improvement over the 1957 value was considered necessary. One was the increasing use being made of very precise electromagnetic distance measuring techniques, both on the Earth's surface and in space, which are based on the speed of light. One example is the study of satellite orbits in which distances are measured by observing the time of flight of laser pulses. The use of the results for investigating the dynamics of satellites involves calculations of acceleration and force, and there must be compatibility with the corresponding terrestrial units of distance. A second example, in the field of measurement standards, was the urgent requirement for a more accurate value of c to enable measurements based on electrostatic quantities to be related to SI units based on the electromagnetic system. Furthermore the cohesion between the SI base units of length and time could be improved. At present they are defined independently; but with a sufficiently accurate knowledge of c they could both be based on the same atomic or molecular transition. This would have the effect thereafter of giving the speed of light a fixed conventional value.

The base units of length and frequency are at present defined in terms of the wavelength and frequency respectively of two atomic transitions, so that the most direct determinations of c are those involving measurements of wavelength and frequency. The speed of light is obtained from the relation $c = f\lambda$, by measuring the frequency and wavelength of some convenient radiation as directly as possible in terms of their respective primary standards. The standard of time interval, the caesium transition near 9 GHz, is the most accurate of all physical standards and can be used with an uncertainty of only a few parts in 10^{13}. The standard of length, the wavelength of orange light corresponding to a krypton-86 transition, can be used with an uncertainty of a few parts in 10^9.

In the measurements described below and in part II, the frequency and wavelength of carbon dioxide laser radiation at 9.3 μm have been determined. The frequency of this radiation was related to that of the frequency standard via two intermediate lasers, HCN and H_2O, and a microwave oscillator. There were close harmonic relationships between the frequencies of the radiations and beat frequencies could be observed and measured at each intermediate stage. To simplify the wavelength measurements, an upconversion technique involving optical mixing in a nonlinear crystal was used. Visible difference-frequency radiation at 679 nm was generated from that of the CO_2 and a visible He-Ne laser, and its wavelength was measured by Fabry–Perot interferometry, with the light from a stabilized He-Ne laser at 633 nm acting as an intermediate wavelength standard. The CO_2 wavelength could be calculated from the result of this measurement, and the method enabled the interferometric observations to be carried out on visible light, by well-understood techniques, and avoided the major systematic effects which would have arisen if the relatively long-wavelength infrared radiation had been used directly. We have reported the progress of this whole experiment (Bradley et al. 1972a,b), and the result was announced in a letter (Blaney et al. 1974).

Our measurement of c is a natural evolution of that undertaken by Froome (1958) with the use of 4 mm microwaves. He achieved an uncertainty for c of ± 3 parts in 10^7, and this represented the uncertainty of the conventional value until 1973. The

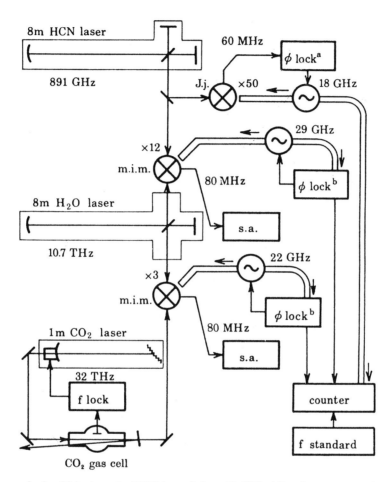

a, lock of klystron to HCN laser (via a 60 MHz i.f. reference crystal).

b, lock of klystron to a harmonic of a 15 MHz quartz crystal oscillator.

FIGURE 1. Laser harmonic-mixing chain to measure the CO_2 frequency. Symbols: s.a., spectrum analyser; ϕ-lock, phase-lock; f-lock, frequency-lock; J.j., Josephson junction; m.i.m., metal-insulator-metal diode.

main source of uncertainty in his experiment was in the determination of the 4 mm wavelength. Apart from uncertainties in calibrating transfer end gauges, which could be reduced by the use of modern laser techniques, the principal contribution was from diffraction of the 4 mm radiation. A comparatively large four-horn interferometer was used in air, requiring separate refractometer measurements to correct the wavelength to vacuum, and a correction was necessary to allow for wave front curvature arising from diffraction at the horn apertures. In our measurement, such problems are reduced or avoided by using a radiation of much shorter wavelength.

The CO_2 laser in the 10 μm region was chosen as the basis for the experiment because it could operate at any one of a large number of closely spaced frequencies. Each offered high power and could be stabilized by reference to a narrow, saturated-absorption feature detected in the fluorescence from a low-pressure CO_2 gas cell (Freed & Javan 1970). The reproducibility of the stabilized laser, a few parts in 10^{10}, safely exceeded the uncertainty of realisation of the length standard, allowing the frequency and wavelength measurements to be made independently. The choice of possible operating frequencies enabled one to be selected for its suitability for measurement via successive stages of harmonic generation and mixing from the microwave through the submillimetre region, where there are only a few laser transitions of adequate power from which to choose transfer oscillators. The high power of the CO_2 laser was also very important for the wavelength measurement, as it made possible the generation of adequate upconverted power at 679 nm (0.3 μW) for an accurate interferometric measurement.

While this experiment was in progress, two other experiments involving frequency and wavelength measurements on laser radiations were completed. The first by Bay, Luther & White (1972) used microwave-modulated light, while the second and more accurate by Evenson et al. (1972) used a chain of harmonically related lasers. The arrangement of this latter experiment was broadly similar to ours, except that it was based on a wavelength measurement of 3.39 μm, and it gave a result of similar accuracy. All three recent results are in good agreement.

In the remainder of part I we describe the frequency measurements of the CO_2 R(12) transition at 32 THz (9.3 μm wavelength) and the stabilized CO_2 laser used. In part II we describe the wavelength determinations and conclude with a discussion of our result for c.

2. Introduction – laser frequency measurement

In the well-established radio and microwave region, i.e. up to about 20 GHz, electronic equipment is readily available that will measure frequencies by direct reference to a rubidium or caesium standard. Digital counting techniques extend to about 1 GHz, covering the usual intermediate frequency (i.f.) region, in which beat frequencies are observed in our experiments. Higher frequencies in the microwave region are measured by generating a harmonic of a lower, countable, frequency to fall near the unknown signal, and counting the resulting difference, or beat, frequency. The process of extending measurement from microwave to submillimetre and infrared laser signals is similar. However, without devices for generating harmonics of a microwave frequency directly to the region of interest, it is necessary to use interjacent transfer oscillators suitably related in frequency, and to construct a chain of harmonic mixing stages.

In practice, to make a measurement in the 10 μm region, near 30 THz, a set of three or four c.w. gas lasers was required such that:

(i) the lowest laser frequency could be measured by direct harmonic mixing with a microwave source;

(ii) the harmonic mixing operations between each laser and the next resulted in relatively low (microwave) beat frequencies; and

(iii) the harmonic orders were practicable with the power levels and nonlinear devices available.

A search from among known c.w. gas laser lines (Knight 1969, 1970) showed a promising chain to consist of the HCN line at 337 μm, the D_2O line at 84 μm, the H_2O line at 28 μm, and the R(10) and R(12) or adjacent CO_2 lines at 9.3 μm. The HCN and H_2O lasers, though not the D_2O, were comparatively powerful and easy to operate. Measurement of the HCN frequency had been demonstrated by Hocker et al. (1967), and the need for the D_2O laser in the chain was eliminated by Evenson et al. (1970), who observed the beat directly between the 12th harmonic of the HCN laser and the H_2O laser with the use of a metal insulator metal (m.i.m.) diode. Mixing between the H_2O and CO_2 lasers had been achieved previously with an m.i.m. diode, but only with pulsed lasers (Daneu et al. 1969).

Our final chain therefore involved the HCN, H_2O and CO_2 lasers with m.i.m. diode mixers, and the scheme is shown in figure 1. The CO_2 laser was stabilized by a method similar to that used by Freed & Javan (1970) while the others were free-running. Klystrons were necessary, most critically as a transfer oscillator at 18 GHz to measure the HCN frequency by 50th-harmonic mixing in a Josephson junction, and to downconvert the HCN/H_2O and H_2O/CO_2 beats to an i.f. near 80 MHz. Phase-lock techniques were used to control and measure the klystron frequencies, but the 80 MHz beats were observed and adjusted manually from observations of spectrum analyser displays, since they were generally too weak to count digitally.

The experimental techniques were developed in previous work involving frequency measurements on an HCN laser (Bradley & Knight 1970; Bradley, Knight & McGee 1971) and on a Lamb-dip stabilized 28 μm H_2O laser (Blaney et al. 1973a). The latter experiment was the first to use a Josephson junction combined with a 'downward' phaselock technique to measure the HCN frequency, and used a computing counter to make and combine the various measurements.

We have announced our result for the CO_2 R(12) frequency with a provisional uncertainty rather less than 1 part in 10^9 (Blaney et al. 1973b). In presenting the results in this paper we obtain the centre frequency of the CO_2 transition to which the laser was locked, by applying a correction of -10 kHz to allow for the background slope of the saturated-absorption feature; but the result for the frequency of the stabilized laser itself from which the speed of light was calculated remains unchanged. The systematic uncertainties have been more closely examined, and reduced.

The following sections are arranged broadly as a description of the frequency-measuring chain (§3) and stabilized CO_2 laser (§4), followed by measurement technique (§5), results and discussion. A separate experiment to test for systematic error in measuring the HCN-laser frequency is described in the appendix.

3. Frequency-measuring chain

The photographs in figure 2, plate 1, give an overall view of the size and layout of the apparatus and of its environment. Our laboratory was on an upper floor with windows on three sides, so that there were difficulties from temperature variation and building vibration. A combination of external sun blinds and extractor fans was used to control the room temperature, and simple vibration isolators were used on various parts of the apparatus.

(a) HCN-laser frequency measurement with a Josephson-junction mixer

The lowest-frequency laser (figure 1) was the HCN one operating at 891 GHz. Both this and the H_2O laser were 8 m long in order to provide sufficient power for harmonic mixing in the point-contact diodes, and were as described by Bradley, Edwards & Knight (1972c). The short-term frequency fluctuations of the HCN laser were reduced by (a) always running the laser with very stable striations in the discharge positive column; (b) running from a current-stabilized power supply (KSM model HVI 5000-2) with less than 0.1 % current ripple; (c) mounting the laser bench on antivibration supports, and (d) reducing the vacuum-pump vibration reaching the laser cavity by firmly anchoring the pump line in a container of sand. With these precautions, the HCN-laser linewidth was about 10 kHz and the laser operated in a single TEM_{00} mode. The longer-term drifts of output frequency and amplitude were reduced by using invar cavity spacers and by working when the ambient temperature drift was less than about ± 0.1 °C/h. The corresponding frequency drift rates of the HCN and H_2O lasers were within the range ± 2 parts in 10^{10} per second and these lasers were not actively stabilized.

Most of the output power from the HCN laser (typically 50 mW) was employed in the harmonic mixing with the H_2O laser, see §3b. However, about 5 % of the HCN laser power, diverted from the main beam with a Melinex beam splitter, was used to monitor the frequency. This was accomplished by generating the 50th harmonic of a klystron frequency f_{k18} (near 18 GHz) and mixing this with the laser frequency f_{HCN} in a Josephson-junction harmonic mixer, the resulting i.f. (at 60 MHz) Δf_{60} being used to phase-lock the klystron frequency to that of the laser (see figure 4).

Description of plate 1

FIGURE 2. Photographs of the apparatus. (a) Looking south: the lasers and H_2O/CO_2 mixing bench. From upper right the lasers are: 8 m HCN (against window), 8-m H_2O with evacuated output arms, 4 m H_2O (not used) and stabilised CO_2 laser. (b) Looking north: HCN/H_2O mixing bench (centre) and electronics racks. The HCN output and beam splitter to the Josephson mixer (in the helium dewar) are at the left.

FIGURE 2. For description see facing page.

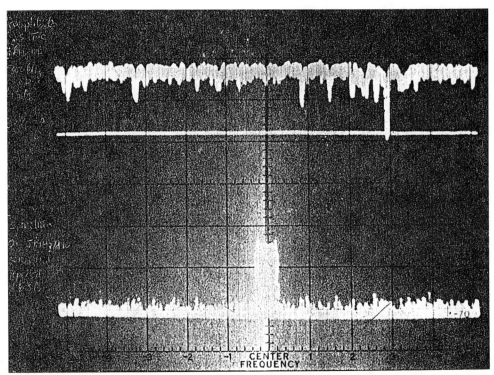

Figure 3. $12f_{HCN}$ against f_{H_2O} beat signal. *Lower trace*, spectrum analyser display similar to that seen by the operator; about six traces stored at intervals over a few seconds with timebase, dispersion and bandwidth of 10 ms/div., 0.5 MHz/div. and 30 kHz and a vertical axis of 10 dB/div. Note the 'square' shape, that the width exceeds the analyser bandwidth for optimum signal/noise ratio, and that the latter was here fairly good, about 14 dB. *Upper trace*, time variation of the beat signal, displayed at 10 ms/div., with 10 dB/div. vertically as before (shifted baseline). Note the quite large amplitude fluctuations, typically as much as 10 dB. The receiver bandwidths were: 300 kHz i.f. and 10 kHz video.

The frequencies are related by:

$$f_{\text{HCN}} = 50 f_{k18} \pm \Delta f_{60}. \tag{1}$$

The klystron frequency was counted as described in §3e. The net result was that counts of the laser frequency could be made continually. The averaging time of individual counts was normally set at 1 s, a period for which the HCN r.m.s. frequency deviation was near a minimum, of about 5 parts in 10^{10}. In this paper 'r.m.s. frequency deviation' refers to the square root of the two-sample Allan variance (Allan 1966), measured with the computing counter described in §3e.

The technique of phase-locking klystrons to the HCN laser has been described in earlier publications (Blaney et al. 1971; Blaney & Knight 1973) and was devised to avoid problems with multiplied phase noise that arise if the klystron is locked to a quartz-crystal harmonic in the conventional manner (Knight 1971). The Josephson junctions used in the present experiments were of a niobium pre-set point-contact type, operating in a liquid-helium storage vessel (Blaney 1971). The experimental details of the use of Josephson junctions in these phase-locking experiments have been described by Blaney & Knight (1973). Both laser and klystron radiations were directed onto the junction via a 'light-pipe' arrangement, while the 60 MHz i.f. signal was amplified by a room-temperature amplifier (gain 45 dB, noise figure 2.5 dB) which was connected to the junction by a miniature co-axial cable. The amplified i.f. was fed to a phase-lock stabilizer which compared the i.f. phase with that of a quartz-crystal reference oscillator. Signals proportional to the phase error were fed to the klystron's reflector, thus completing the phase-lock loop. The stabilizer was a commercial type (Microwave Systems PLS-60), having a modified lead-lag filter to compensate for the frequency multiplication in the loop, as described by Blaney & Knight (1973). The operating characteristics of the Josephson harmonic mixers have been described in a series of publications (Blaney & Knight 1973, 1974a,b). With the arrangement used, it was usually necessary, every 10 min or so, to make small manual adjustments to the microwave power level or d.c. bias current at the Josephson mixer in order to maintain the amplitude of the i.f. signal at a level sufficient for continuous stable phase locking.

The use of Josephson junctions in high-accuracy frequency measurement work is relatively novel, and there is at present no detailed understanding of the harmonic mixing mechanism, at least in point contacts. Thus it is necessary to consider whether the unique properties of these devices could give rise to systematic frequency errors when used in the way described here. For example, intrinsic in Josephson systems is the existence of internal oscillations whose frequency is dependent on the voltage bias (Josephson 1962, 1964). However, in all the harmonic mixing experiments carried out at the N.P.L., no i.f. signals corresponding to these internally generated frequencies have ever been observed. A similar result has been reported by McDonald et al. (1971) following experiments at N.B.S. Boulder. This is not surprising, as free-running Josephson oscillations are likely to have a relatively large linewidth and be of unstable frequency owing to noise and voltage fluctuations in the system. We have also carried out experiments in which the possible existence of systematic frequency offsets could be investigated. The discussion in the appendix includes a comparison of frequency multiplication in a Josephson junction with that in an m.i.m. diode. The results showed no significant difference between the measurements by the two methods at the 1.2 parts in 10^{10} uncertainty level (70 % confidence) of the comparison. We thus conclude that Josephson junctions, as used here, do not exhibit peculiar intrinsic properties likely to introduce significant frequency errors.

(b) *HCN-laser/H$_2$O-laser mixing stage*

The twelfth harmonic generation and HCN/H$_2$O mixing processes were carried out by means of a room-temperature m.i.m. (metal-insulator-metal) diode. These devices have been fairly extensively studied since the first use by Daneu *et al.* (1969) of their properties as harmonic generators in the infrared. Our device consisted of a tungsten whisker 25 μm in diameter and 2 mm long in light contact with a polished nickel post 2 mm in diameter. The whisker had a typical tip radius of 0.05 μm as measured with an electron microscope. This was obtained by electrolysis in a solution of KOH, and the contact pressure of the junction between the two metals could be varied by means of a lever movement in conjunction with a micrometer screw. The diode was mounted on an X–Y–Z translation stage, which could be rotated about a vertical axis, with the whisker in the horizontal plane, in which the laser radiation was polarised. The i.f. output was coupled from the whisker on a 50 Ω coaxial line. Details of the m.i.m. diodes have already been published (Bradley *et al.* 1972*b,c*; Bradley & Edwards 1973).

The diode was placed between the output windows of the HCN and H$_2$O lasers, and the laser radiations were focused on to the junction with respectively a polythene lens of 70 mm focal length and an off-axis paraboloid mirror of 120 mm focal length. An open-ended waveguide, fed by a klystron, was situated about 10 mm above the junction and served to irradiate the diode with the 29 GHz radiation necessary to downconvert the i.f. to 80 MHz. The frequencies are related by:

$$f_{H_2O} = 12 f_{HCN} + f_{k29} \pm \Delta f_{80}, \qquad (2)$$

where f_{H_2O}, f_{k29} and Δf_{80} are the H$_2$O, 29 GHz and 80 MHz frequencies.

The construction of the H$_2$O laser broadly followed that for the HCN laser and has been briefly described by Bradley *et al.* (1972*c*). Two gold-coated mirrors, one plane and one of 20 m radius, were mounted in aluminium plates maintained 8 m apart by four 25 mm diameter invar rods. The discharge tube was water cooled and the flowing-gas discharge was operated typically at 0.5 A provided by a current-stabilized power supply, as for the HCN laser. Power at 28 μm wavelength was coupled out of the laser by means of a single, 10 μm thick polypropylene or polythene film mounted at 45° to the laser axis. Evacuated tubes were used for the majority of the output beam paths to the mixing diodes, to reduce atmospheric absorption.

The total power obtained at 28 μm from each of the two output ports was estimated to be in the region 30–50 mW. Since the mode spacing, $c/2L$, is about 19 MHz for an 8 m long cavity, as compared with the laser tuning width of approximately 45 MHz, this power was distributed over two or three modes. Usually one mode was centred in the gain profile and emitted the bulk of the power.

Care was taken with the focusing of the radiations and with the diode orientation with respect to the beams, particularly that from the HCN laser, in order to optimize the effective diode cross section. It has been demonstrated experimentally that the whisker acts as a long antenna at these wavelengths (Matarrese & Evenson 1970), and for a 2 mm long whisker the optimum angles between the Poynting vector and whisker were about 20° and 10° respectively for the HCN and H_2O laser beams. Alignment was aided by observing the focal spots of these beams with heat-sensitive liquid-crystal sheet or Thermofax paper.

The i.f. signal at about 80 MHz was passed through an amplifier (40 dB gain, 4 dB noise figure) and displayed on a spectrum analyser (Hewlett Packard 8553B, 8552B, 141T). The beat note seen on the analyser display (figure 3, plate 2) had a signal to noise ratio that was typically 12 dB, with a maximum of 24 dB. The noise level was dominated by noise generated by the diode due to the incident HCN radiation; the mechanism for this process is not clear and the noise was sometimes unstable with a period of the order of $\frac{1}{2}$ s. Such instabilities were usually characteristic of diodes with short lifetimes (up to a few hours) and more normally a whisker would last for a day or two until it was blunted enough to need replacement. The HCN/H_2O beat signal was initially difficult to obtain, although after the first success the process became relatively easy. An important indicator of whether a diode would produce a beat signal was the presence on the spectrum analyser of the wide-band noise when the diode was irradiated by the HCN laser. Typically this noise level would be 10–15 dB above the ambient noise from the diode when not irradiated.

The width of the beat signal on the spectrum analyser was in the region of 100–150 kHz, when the rate of scan was such as to take about 1 s to cross the signal. Most of this was derived from acoustic-frequency fluctuations in the frequencies of the two lasers, leading to the square-topped signals observed in these experiments (see figure 3, plate 2).

(c) H_2O-laser/CO_2-laser mixing stage

The procedure in the H_2O-laser/CO_2-laser mixing stage was very similar to that just described. An m.i.m. diode, situated between the second output window of the H_2O laser and the CO_2 laser, was used to mix the laser radiations. The resulting i.f. was downconverted to 80 MHz by the use of a 22 GHz klystron whose waveguide terminated just above the diode junction. The frequencies are related by:

$$f'_{R12} = 3f_{H_2O} + f_{k22} \pm \Delta f'_{80}, \tag{3}$$

where f'_{R12}, f_{k22} and $\Delta f'_{80}$ are the CO_2-laser, 22 GHz and 80 MHz frequencies. (A prime is applied to the CO_2-laser frequency to distinguish it from the centre frequency of the transition to which it is locked, f_{R12}.) Focusing of the H_2O and CO_2 laser beams was achieved by the use of a 120 mm focal-length off-axis paraboloid mirror and a 100 mm focal-length barium fluoride lens respectively, and the nominal powers incident on the diode were 40 and 150 mW.

The 80 MHz i.f. signal was amplified and displayed on a second spectrum analyser. After optimizing the focusing parameters, klystron power, etc., signal-to-noise ratios up to 35 dB were obtained. The widths of the beat spectra were in the region 300–500 kHz, as observed with a slow analyser scan, when the CO_2 laser was free running, but when the CO_2 laser was locked to its external CO_2 cell, the beat width was increased by the peak-to-peak value of the frequency modulation applied to the laser.

It has already been noted that the H_2O laser oscillated in two or three modes simultaneously, and as a result of the various combinations possible between these modes and the CO_2 laser frequency, five or six beat signals spaced by the 19 MHz H_2O intermode frequency and of irregular amplitude could be found on the spectrum analyser display. To identify which beat in fact corresponded to the 3rd harmonic of the strongest H_2O mode which was being measured in the H_2O/HCN stage a preliminary experiment was done. The H_2O laser was made to run in a single mode by reducing the current and hence the gain, leaving only one beat note on the screen. Although the HCN/H_2O beat could not be seen under these conditions, we measured the CO_2 frequency to about ± 1 MHz by setting the H_2O laser to the peak of the power-output curve, observed with a Golay detector. With the laser operating under normal conditions the correct beat signal, corresponding to this preliminary measurement, could then be unambiguously selected.

(d) *Klystron phase-locking*

The klystrons were of the conventional reflex type that can be tuned over about 0.1 % of the frequency by varying the reflector voltage. They were water cooled by epoxy-bonded pipes, to improve the temperature stability, and were run from well-stabilized d.c. supplies. Their stability was further improved by mounting them on vibration isolators. The free-running r.m.s. frequency deviation for 1 s averaging was about 1 part in 10^7 (Knight & Brown 1973).

Phase locking the 18 GHz klystron to the HCN laser has been discussed in §3a. The klystrons used to downconvert the inter-laser beats were stabilized in the conventional manner by phase locking to harmonics of 15 MHz quartz-crystal oscillators. These oscillators were tuned manually to adjust the spectrum-analyser beats to 80 MHz. Commercial equipment (Microwave Systems Inc.) was used for stabilization in this way, but its operation was carefully checked to ensure the absence of spurious sidebands either from feedback-loop oscillation or from oscillation in the multipliers generating the quartz-crystal harmonic.

The klystron frequency f_k was related to the 15 MHz quartz-crystal frequency f_{15} by:

$$f_k = 12nf_{15} \pm \Delta f_{40}, \qquad (4)$$

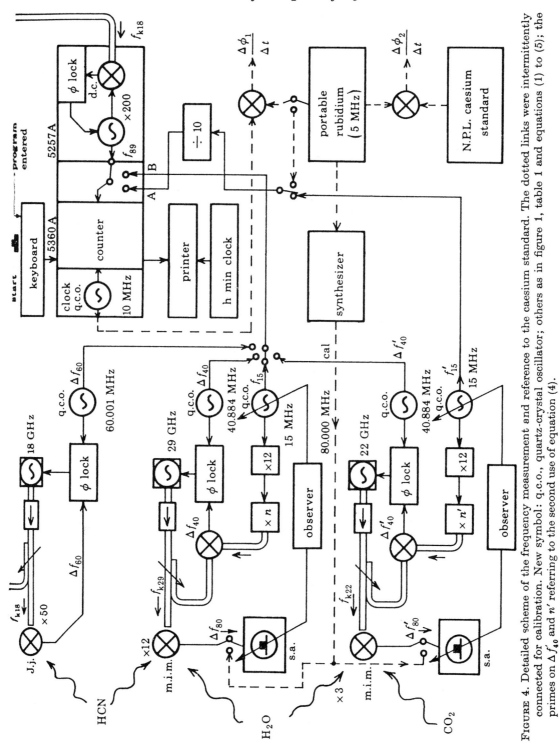

FIGURE 4. Detailed scheme of the frequency measurement and reference to the caesium standard. The dotted links were intermittently connected for calibration. New symbol: q.c.o., quartz-crystal oscillator; others as in figure 1, table 1 and equations (1) to (5); the primes on $\Delta f'_{40}$ and n' referring to the second use of equation (4).

where n is a harmonic order, referring to one of a 'comb' of microwave harmonics spaced by 180 MHz, and Δf_{40} is the 'fixed' frequency of an i.f.-reference quartz-crystal oscillator, near 40 MHz. The schemes are shown in figure 4.

When locked, the klystron drift rate and spectral width were controlled by those of f_{15} and were so small (Knight 1971) as to make negligible contributions to the uncertainties of the HCN/H_2O and H_2O/CO_2 beat measurements. Drift in an i.f.-reference crystal was only significant in the lock of the 18 GHz klystron to the HCN laser, i.e. in Δf_{60}, so that it was necessary to monitor this regularly.

TABLE 1. KLYSTRON AND LASER OPERATING FREQUENCIES

oscillator	symbol	frequency†/MHz		laser beat freq./MHz
		lower sideband	upper sideband	
CO_2 laser	f'_{R12}		32 176 079	80
22 GHz klystron	f_{k22}	21 792‡	21 952	
H_2O laser	f_{H_2O}		10 718 069 ± 8	80
29 GHz klystron	f_{k29}	28 869	29 029	
HCN laser	f_{HCN}		890 760 ± 1	60
18 GHz klystron	f_{k18}	17 814.00	17 816.40	
transfer osc.	f_{89}	89.0700	89.0820	

† The laser frequencies are rounded to the nearest MHz. The HCN and H_2O frequencies could drift within the limits shown, and the klystron frequencies were varied by as much as ±20 MHz to compensate. The frequencies are related by equations (1) to (6).

‡ The centre klystron frequencies, corresponding to the centre HCN and H_2O frequencies, are given.

Unambiguous checks of the klystron frequency were made during measurements, with wavemeters calibrated to about 0.01 %, i.e. 3 MHz in 30 GHz. These identified the harmonics and sidebands of the references to which the klystrons were locked, but could not guard against an offset resulting from oscillation in a feedback loop. Checks against this were provided by i.f. frequency discriminators in the stabilizers themselves, any offset in the locked i.f. being equal to an integral multiple of the oscillation frequency, which in such loops tends to be of the same order as the loop bandwidth, in this case about 100 kHz. Occurrence of such an offset would also have been observed as a 'jump' in the CO_2 frequency-measurement results.

(e) Relation of frequencies to the caesium standard

The detailed scheme of the frequency measurements is shown in figure 4 and the various laser and klystron frequency values are listed, according to the sidebands used, in table 1.

The measurements that were made in quick succession with the electronic counting apparatus comprised:

(i) f_{k18}, the frequency of the 18 GHz klystron phase locked to the HCN laser (equation (1)),

(ii) f_{15}, the 15 MHz crystal frequency to which f_{k29} (equation (2)) was locked (according to equation (4)), and similarly

(iii) f'_{15} the 15 MHz crystal frequency to which f_{k22} (equation (3)) was locked.

The three signals were applied to separate inputs of a computing counter having a nominal accuracy, for 1 s averaging, of 1 part in 10^9. For the 18 GHz signal a plug-in transfer oscillator was used that phase locked an internal oscillator at an exact subharmonic of the input signal, in our case at a frequency f_{89}, near 89 MHz, such that:

$$f_{k18} = 200 f_{89}. \qquad (5)$$

The harmonic order 200 was chosen for ease of reading f_{HCN}, since then from equation (1):

$$f_{HCN} = 10\,000 f_{89} \pm \Delta f_{60}. \qquad (6)$$

f_{89} is the frequency which was counted, and the uncertainty in the measurement had to be smaller than that required in the overall CO_2 measurement. (A front-panel monitor point allowed the state of lock of f_{89} to f_{k18} to be checked.) The measurement of f_{15} and f'_{15} was much less critical.

Descriptions of the computing counter, its keyboard and the 18 GHz transfer oscillator (Hewlett Packard models 5360A, 5375A and 5257A) have been given in the Hewlett Packard Journal in the issues of May 1969, March 1970 and February 1968 respectively. With a suitably written program the computing counter was instructed to read the inputs (i) to (iii) above, in turn, averaging each for 1 s, and to calculate f'_{R12} from equations (6), (2), (3) and (4). One manual operation was required at the end of the stored program to complete the calculation because of the computing counter's limited constant storage. The CO_2 frequency and a number related to the H_2O frequency were calculated and printed in about 3 s every time the experimenters were ready (see §6). To allow measurement of drift and retrospective checks of results against, for example, temperature and humidity records, the time of each record was also printed. The apparatus thus gave immediate information both on the state of the CO_2 measurement and the operation of the free-running lasers used as transfer oscillators. The choice of 1 s averaging time was a good compromise between laser stability and counter accuracy.

All measurements were referred in the first instance to the internal 10 MHz quartz-crystal clock of the computing counter. This oscillator, undisturbed and with the temperature-control oven kept permanently on, showed an ageing rate of only a few parts in 10^{10} per month. It was calibrated daily by comparison with a portable rubidium standard, either by injecting a rubidium signal into the counter or by an external phase-drift comparison. The rubidium standard was compared with the N.P.L. caesium standard before and after each run, the difference being typically ± 1 part in 10^{11}. The correction made afterwards to f'_{R12} for the computing-counter clock offset from the caesium value varied between 1 and 4 parts in 10^{10} during the experiment. Our rubidium standard had insufficient short-term stability simply to substitute it in place of the clock of the computing counter.

Sources of possible inaccuracy in the computing counter were checked. The

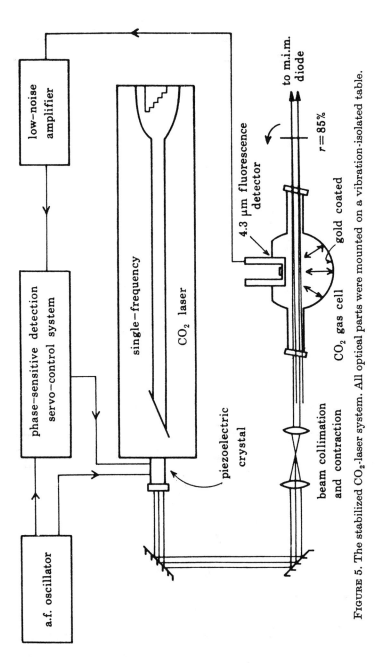

Figure 5. The stabilized CO_2-laser system. All optical parts were mounted on a vibration-isolated table.

digital arithmetic used in the program was compared with that of a double-length arithmetic machine and showed differences, at the level expected due to rounding, of about 3 parts in 10^{11}. In addition, the computing counter uses a phase-interpolation method to increase its accuracy beyond the ± 1 part in 10^7 that would result from counting the number of clock zero crossings in 1 s. The interpolator accuracy was tested both by the instrument self-check mode, and by applying a 500 kHz signal from the rubidium standard. A histogram of 290 one-second averages was compared with the reading for a 30 s counting time, in which the interpolator error is 30 times smaller (nominally 3 parts in 10^{11}). While taking successive readings for the histogram the relative phase of the rubidium signal moved 0.7 of a cycle referred to the clock. The histogram was approximately Gaussian with a standard deviation of 2.3 parts in 10^{10} and standard error of the mean of 1.6 parts in 10^{11}. Its mean agreed well with that of the 30 s average, showing no observable systematic offset arising from the interpolation process. Our conclusion is that at the time of the test, shortly after completing the measurements, the interpolator errors were predominantly random, at the level of 2 parts in 10^{10} for individual 1 s counts. However, in view of the nature of the counting circuits it may have been fortuitous that a systematic offset much less than 1 part in 10^{10} was observed.

The systematic uncertainty of 2 parts in 10^{10} given in table 4 as 'counting errors' comprises approximately equal allowances for the latter effect and for possible drift of the computing-counter clock between calibrations, with a small addition for machine arithmetic error. The further allowance of 2 parts in 10^{10} for 'room temperature drift' derives from a possible net drift of the HCN laser frequency, §3a, in the time delay of the order of 1 s between the operators' setting the HCN/H_2O and H_2O/CO_2 beats and counting the HCN frequency, as described in §5.

4. Stabilized CO_2 lasers

The stabilized CO_2 laser system shown in figure 5 (Woods & Jolliffe 1976) was based on a 1 m sealed CO_2 laser with grating tuning, with part of the output used to illuminate a low-pressure CO_2 gas cell with oppositely-travelling waves. A central tuning dip in the 4.3 μm fluorescence that resulted from saturated absorption in the gas cell (Freed & Javan 1970) provided a reference free from the first-order Doppler effect, and the laser was locked to it with a first-derivative servo. Simple stabilization of the CO_2 laser frequency to the power-output peak would have been unsatisfactory, because the servo frequency scan would have excessively broadened the spectrum, and in addition the centre frequency itself would have depended on, for example, the gas fill and age of the gain tube.

The saturated-absorption technique of Freed & Javan was modified by placing the gas cell outside the laser cavity, rather than inside. This allowed us to investigate the accuracy of stabilization to the centre frequency by varying cell parameters such as gas pressure, laser radiation intensity and wavefront curvature. One of the practical difficulties of this experiment was to obtain single-frequency and stable operation of the laser with sufficient power, since approximately 1.5 W each was needed for stabilization and the wavelength measurement, while a sealed laser was necessary for stability and the R(12) line of the 9.3 μm band was of comparatively low gain. It was also necessary to get the best possible signal to noise ratio in detecting the 4.3 μm fluorescence, particularly since the frequency scan used for locking to the dip had to be kept smaller than the optimum to avoid excessively broadening the spectrum for the frequency measurement.

(a) Description of CO_2 laser

The laser cavity was spaced by four 19 mm diameter fused-silica rods encased elastically in a rigid aluminium box. This design gives a low coefficient of thermal expansion, high thermal inertia and good damping of vibrations in the spacing rods

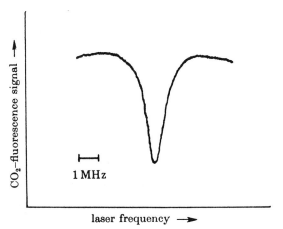

FIGURE 6. Saturated-absorption dip in the CO_2 fluorescence. The dip size is approximately 9 % of the peak fluorescence value, and the width is 1.06 MHz (f.w.h.m.) for a 3.2 Pa cell pressure. (20 s scan, time constant 100 ms.) The slight asymmetry (see the higher 'shoulder' on the low-frequency side) arises from the same effect (the shift of the fluorescence-profile centre from that of dip) that gives rise to the -10 kHz correction discussed in §4b.

(Woods & Jolliffe 1976). One end reflector was a Littrow-mounted gold-coated blazed diffraction grating (PTR Optics Inc. Type SF300) to provide wavelength selection, and the other was a 5 m radius concave germanium mirror of 85 % reflectivity that provided the output. This was mounted on a piezoelectric ceramic transducer (PZT) to control the frequency. The gain tube was of fused silica and of 8 mm bore, with one Brewster window (of KCl) at the mirror end. Details of the gas fill, cathode and discharge-tube design are as given by Woods & Jolliffe. The discharge length was 0.8 m and it was run from a current-stabilized supply (0.01 % ripple and noise), typically at 12 mA, 12 kV. A mixture of water and glycol, temperature controlled between 0 and 4 °C, was circulated through a cooling jacket on the gain tube in order to increase the power output on the R(12) line.

The absorption cell normally used was about 150 mm long, with a central spherical region about 80 mm diameter having the 4.3 μm fluorescence detector mounted near the centre of curvature (see figure 5). The detector was a liquid-nitrogen cooled indium antimonide type (Mullard RPY35). The surface of the pyrex bulb was gold coated to reflect radiation, both to maximize collection of the fluorescent radiation and to shield the detector thermally by returning its own radiation. Angled KCl or BaF_2 windows transmitted the CO_2 radiation. This was concentrated in a beam of

TABLE 2. SYSTEMATIC UNCERTAINTIES OF STABILIZED CO_2-LASER FREQUENCY

source	uncertainty ± kHz
1. Reproducibility: affecting f'_{R12} and f_{R12}. Variation of:	
(i) electronic offset	0.9
(ii) frequency scan (± 33 %)	0.8
(iii) laser power (affects %-size of dip)	0.8
(iv) beam size and parallelism in gas cell	1.0
(v) cell pressure (± 1 Pa)	2.5
(vi) possible power shift of reference transition centre (1–2 W)	4.0
(vii) further 'day to day' effects (e.g. feedback into laser)	4.0
quadrature sum	6.4
2. Affecting f_{R12}: unperturbed transition-centre ($P = 4$ Pa) value	
(i) offset correction uncertainty	5.3
(ii) possible power shift of transition centre	4.0
quadrature sum	6.6

about 2 mm diameter on the cell axis and was returned through the cell by an external 85%-reflecting germanium plane mirror. The transmitted beam was used for the frequency measurement. The returned beam was deflected by about 4 mrad to avoid, as far as possible, any optical feedback into the laser. (Feedback from, for example, a damaged KCl window was found to cause shifts in the locked frequency of order ± 20 kHz.) The CO_2 gas in the cell was held at a pressure of about 4 Pa (30 mTorr) and was flowed slowly through it to guard against any effect arising from leaks. For the frequency measurements the laser power was in the range 1.2–2.8 W, giving a saturated-absorption dip of between 8 and 11 % of the peak height of the fluorescence signal, and a width (f.w.h.m.) typically of 1.3 MHz for a gas pressure of 4 Pa. This was 4 parts in 10^8 of the frequency and known contributions include those from pressure (320 kHz), power (200 kHz), and transit time (60 kHz). A dip profile is shown in figure 6.

The servocontrol electronic system was similar to that described by Shotton & Rowley (1975). The CO_2 laser frequency was scanned sinusoidally with an amplitude of about 450 kHz peak to peak at a frequency near 300 Hz, by a signal applied to the piezoelectric mirror mount. The resulting intensity modulation of the fluorescence signal was taken via a battery-powered, low-noise amplifier and coherent filter to a phase-sensitive detector. It was then fed back to the laser in the appropriate phase via an integrating filter in order to hold the laser frequency to the position of zero slope at the bottom of the fluorescence dip. The integrating servo was adjusted to have unity gain at about 8 Hz with a 6 or 9 dB/octave roll-off (on different occasions) for best results. The noise in the first-derivative servo signal was about 50 mV peak-to-peak in a 30 Hz bandwidth, compared with a ± 10 V range to the inflexion points of the Lorentzian-dip profile, which were separated by about 0.75 MHz.

The r.m.s. frequency deviation of the free-running laser reached a minimum of 1.4 parts in 10^{10} for a 0.1 s averaging time τ before the curve turned up from the effect of thermal drift. When stabilized, the r.m.s. frequency deviation was about 2.5 parts in 10^{10} for a 1 s averaging time and proportional to $\tau^{-\frac{1}{2}}$ between 0.1 and 10 s (Woods & Jolliffe 1976).

(b) CO_2-laser frequency reproducibility and accuracy

The effects of changes in system parameters on the frequency of the stabilized laser were investigated experimentally as far as possible, in order to assess:

(i) the reproducibility of the stabilized laser frequency f'_{R12}, and

(ii) its mean offset from the centre of the unperturbed transition in the absorbing gas f_{R12}, and the uncertainty in its correction.

Some experiments were carried out with a pair of similar lasers stabilized on the adjoining R(12) and R(14) transitions, by measuring accurately the 41 GHz difference frequency. This avoided the problem of spectrum overlap that occurs in beating together two lasers locked to the same transition, and permitted observation of the shifts due to changes in various parameters, at the level of about 1 part in 10^{10}, i.e. 3 kHz. Other effects, such as offset in the first-derivative lock position arising from servo-signal background slope, were examined by tests on individual systems.

A list of contributions to the systematic uncertainty thought appropriate to these frequency measurements is shown in table 2: items in part 1 relate to the laser frequency f'_{R12} and those in part 2 are the additional uncertainties which arise in deducing the unperturbed frequency f_{R12}. The first five items of part 1 arise from tolerances in separable system parameters and are relatively small. The largest, (v), corresponds to the pressure variation of ± 1 Pa in the reference cell, the shift being -2.65 ± 0.45 kHz/Pa (Woods & Jolliffe 1976). Item (vi) allows for possible power shift of the reference-transition centre, over the range of intensities used, whereas item (ii) of part 2 from the same effect is an allowance for shift from zero interrogating power, both allowances representing a limit of resolution in experimental tests. Item (vii) of part 1 is an allowance for long-term effects in the stabilized-laser system, such as variations in stray optical feedback into the laser with different alignments of the optics, see §4a.

To obtain f_{R12}, the unperturbed reference-transition, a correction was applied to the laser frequency f'_{R12}. This is necessary because there is frequency shift between the absorption profile in the stabilization cell and the emission profile of the laser, due mainly to their different pressures (Freed & Javan 1970). The 4.3 μm fluorescence curve is a convolution of the absorption profile and the shifted laser profile, whereas the dip is centred on the absorption profile alone. The dip in the 4.3 μm fluorescence is thus not centred on its fluorescence profile, and the resulting background slope affects the first-derivative lock point. For transitions between the same vibration bands at least, this shift in all such stabilized lasers will be in the same sense.

Studies of the fluorescence signal both with and without the returned CO_2 beam blanked off gave the following correction to f_{R12}' to obtain f_{R12}:

$$-10 \pm 5.3 \text{ kHz}.$$

The uncertainty corresponds to possible variations in this correction throughout the experiment and appears in table 2 as (i) of part 2. This and the other uncertainties in table 2 should be considered equivalent to 70%-confidence random uncertainties. The respective totals amount to 2.0 and 2.1 parts in 10^{10} and are carried forward to table 4.

5. MEASUREMENT TECHNIQUE

The problem involved in making a measurement of the CO_2 laser frequency, assuming that all parts of the apparatus are functioning correctly, may be understood by considering the observations which have to be made (figure 4). These comprise:

(i) the three measurements of klystron frequencies made by the computing counter, and

(ii) the two measurements of the centre frequencies of the two beats seen on the spectrum analysers.

Also, since the two transfer lasers are not actively stabilized, it is necessary to make these observations in a time short compared with the time over which significant drift may occur in the laser frequencies.

As well as the above observations, which need to be repeated for every measurement of the CO_2 frequency, other parameters need to be checked at regular intervals. These include measurements of the crystal frequencies used as i.f. references in the boxes used for phase-locking the klystrons; a comparison of the reference crystal of the 5360A computing counter against a local rubidium standard clock; observations of the sidebands to which the klystrons are tuned; checks of the klystron frequencies with cavity wavemeters, and periodic checks of the stabilization loop of the CO_2 laser.

Apart from all these factors there are of course other sources of systematic error which may exist and affect the result in some way, and any parameter which we considered might cause such an error was varied during the course of the measurements, as discussed in detail below.

Returning to the actual measurement technique, in each measurement of the CO_2 laser frequency the following procedure was adopted. Two people were in charge of the two beat notes displayed on the respective spectrum analysers, each of which was near the lock-box controlling the klystron contributing to that particular beat note. By adjusting the frequencies of the crystals to which the klystrons were locked, they could position the beat notes to lie symmetrically across a calibrated (80 MHz), vertical, graticule line on the spectrum analyser display, as in figure 3. When both observers were satisfied that they had optimized the position of their particular signal, a third person activated the 5360A computing counter which, in sequence, counted the frequency of the 18 GHz klystron locked to the HCN laser, and the two crystal frequencies controlling the other klystrons. During this counting procedure, no controls were altered. The counter then computed the CO_2 laser frequency and, as a check, the H_2O frequency from this data, and the result was automatically printed.

TABLE 3. SUMMARY OF THE RESULTS

date	no. of readings	mean† kHz	standard dev.‡ kHz	standard error of the mean‡/kHz
10 Apr. 1973	40	460	90	14
12 Apr. 1973	120	487	67	6.1
13 Apr. 1973	120	481	51	4.7
7 June 1973	100	488	62	6.2
8 June 1973	100	481	63	6.3
overall	**480**	**482**	**63.7**	**2.9**

† The mean of: $(f'_{R12} - 32\,176\,079\,000)/\text{kHz}$.
‡ For results taken individually, not in sets of ten.

The results (table 3) were taken in sets of ten, and within each set no parameters of the experiment were altered apart from a deliberate displacement and resetting of the beat signals on to the calibrated graticule lines. Between each set of ten readings however, parameters were altered and also the calibrations of the graticule lines were checked by feeding 80 MHz signals into the spectrum analysers from a frequency synthesiser. Among parameters which were altered from set to set were: variables which might affect the reproducibility of the CO_2 laser, including modulation level; pressure in the CO_2 absorption cell; the sweep rate and bandwidth of the spectrum analysers; the klystron sidebands (table 1 and equation (4)); the observers controlling the positions of the beat notes. In fact, within the resolution achieved by our results, no effects could be observed due to alteration of any of these variables, apart from deviations caused by some extreme and untypical values of parameters associated with the CO_2 laser, such as very large frequency-modulation levels. Because of this, we can quote the typical values of parameters of the two beat notes. For the HCN/H_2O laser beat these were: signal-to-noise ratio 8–14 dB (on 10 dB/div. display); width (full, at -3 dB) 120–200 kHz; spectrum analyser bandwidth, dispersion and scan rate 10 kHz, 0.2 MHz/div. and 50 ms/div. (with free-running timebase) respectively. In the case of the H_2O/CO_2' signal these parameters were, in the same order, 10–15 dB; 500–700 kHz; 10 kHz; 0.5 MHz/div.; 50 ms/div. In the latter case, the signal width includes the contribution due to the

CO_2 laser frequency modulation, which is additive to the width obtained with free-running lasers.

It is worth noting that, because of the rectangular nature of the displayed beat signals, no great advantage is obtained by having a signal-to-noise ratio greatly in excess of that necessary to see the edges of the signal clearly, about 6 dB. At times, during measurements on 7 and 8 June all three laser beats (including that by which the 18 GHz klystron was locked to the HCN laser) were only just observable, yet no difference or shift was seen in the results (table 3). Also, it seems probable, and our experience tends to confirm it, that the eye centres the beat at a position where

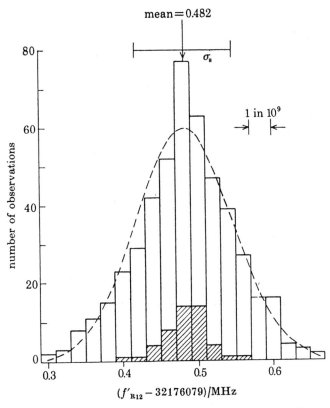

FIGURE 7. Histogram of all 480 individual results, and (shaded) when grouped into the 48 measurement sets. The dotted curve is a fitted Gaussian distribution with a standard deviation σ_s of 64 kHz and the indicated mean.

the calibrated graticule line falls equidistant from the edges of the beat signal. Because of the nature of the display on a spectrum analyser, this 'space-averaged' centre of gravity of the displayed signal can be different from the time-averaged signal frequency if there is asymmetry in the frequency fluctuations of one of the lasers involved. This effect cancels out for the H_2O laser, but not for the HCN laser, because time-averaging is used in the case of the 18 GHz/HCN beat and 'space-averaging' with the HCN/H$_2$O and H$_2$O/CO$_2$ signals. This possible asymmetry correction has been investigated and is discussed more fully in the appendix.

6. Analysis of results as random data

Results on the frequency of the stabilized CO$_2$ laser were obtained on 5 days over a period of 9 weeks, and table 3 lists, on a daily basis, some basic statistical parameters of the results. These parameters are calculated from the values printed out by the computing counter for the CO$_2$ frequency together with a number of applied corrections, calculated from (i) the differences from nominal of the three 'fixed' i.f. frequencies used for phase locking the klystrons, and (ii) the calibration, by means of the rubidium clock, of the reference crystal in the computing counter. The corrections were found to be sensibly constant over the period of a day or so and they were measured only once or twice a day during the period of a run, that for Δf_{60} being the most critical (§ 3d).

We have already noted (§ 5) that no correlation could be found between the results and the adjustable parameters of the measurement system. On this basis, it should be possible to treat all 480 results as random samples of the CO$_2$ laser frequency. This deduction is largely confirmed by all the statistical tests (Campion, Burns & Williams 1973) that have been applied to the data, including t-tests on the deviations of the means of the samples of ten results from the overall mean, and χ^2 tests on both the variations of the variances of the sets of ten and the 'goodness-of-fit' of the results to a normal distribution. Some doubt arose as to the validity of the first day's results, which have a much wider internal scatter than those of the other days and in particular deviate widely from a normal distribution. However, analysis of the individual four sets of data obtained on that day showed that no individual set was statistically improbable at a 5 % level. In any case rejection of this day's results would give only a 2 kHz shift in the overall mean.

Observation of table 3 shows that there is good consistency between the means of the 5 days' results and the standard errors of the means for each day.

From all 480 results, the mean obtained for the frequency of the stabilized laser operating on the $R(12)$ transition of CO$_2$ is:

$$f'_{R12} = 32\ 176\ 079\ 482\ \text{kHz}$$

with a standard deviation for a single reading (σ_s, 70 % confidence level) of 64 kHz (2 parts in 10^9). In figure 7 we have plotted a histogram of the results, together with a curve calculated assuming a normal distribution with the mean and standard deviation above. A measure of the agreement between the calculated and observed distributions is provided by the χ^2 test. The value for χ^2 with 12 degrees of freedom is 14.3, and the theoretical value corresponding to a 5 % probability is 21, so that there are no grounds for rejecting the assumption that our results follow a normal distribution. A histogram of the 48 measurement sets, means of ten results, is also shown for comparison in figure 7, as the shaded area.

The standard error of the mean of all the results may be calculated as $\sigma_s/\sqrt{480} = 2.9\,\text{kHz}$. An alternative approach is to calculate the standard error of the mean of the 48 measurement sets, leading to a figure of 4.2 kHz. For a purely random sample, the expectation values of these two approaches are identical. However, an F-test shows that this difference is significant at the 0.1 % level, indicating that there are correlations between results within individual sets of ten. Because of this we have taken the larger value of the two error estimates, i.e 4.2 kHz, 1.3 parts in 10^{10}, as the more appropriate.

TABLE 4. SYSTEMATIC UNCERTAINTIES

source	± parts in 10^{10}
rubidium standard against caesium	0.3
counting errors	2
room temperature drift	2
HCN laser asymmetry	2
CO_2 laser reproducibility	2
CO_2 laser offset from unperturbed transition centre†	2.1
quadrature sum	4.6

† Comprises 1.7 parts in 10^{10} from the applied $-10\,\text{kHz}$ background-slope correction together with ± 1.3 parts in 10^{10} from possible power shift in the observed transition. It is omitted from the uncertainty of the laser frequency f'_{R12} when used as a transfer oscillator.

7. SYSTEMATIC UNCERTAINTIES

In this experiment the major part of the uncertainty in the result comes from possible systematic effects in the apparatus. The main contributions to the systematic uncertainty are listed in table 4, and are explained more fully in the appropriate parts of the text, mainly §§ 3e, 4b and the appendix. Since our last presentation of the uncertainties in this measurement (Blaney et al. 1975) those appropriate to the CO_2 laser have been re-examined and reduced, with the application of a small correction for offset caused by background-slope in the fluorescence profile, which was determined experimentally to be $-10 \pm 5\,\text{kHz}$.

Our final result for the frequency of the unperturbed R(12) transition in CO_2 at 9.3 μm is therefore:

$$f_{R12} = 32\,176\,079\,472 \pm 15\,\text{kHz}.$$

The total uncertainty, ± 5 parts in 10^{10}, is derived as the sum of the squares of the random and systematic components, 1.3 and 4.6 parts in 10^{10} respectively, and is to be treated as having a 70 % confidence level. This is appropriate to a pressure of 4 Pa, the pressure shift being about $-2.6\,\text{kHz/Pa}$ (§ 4b).

For the purpose of calculating the speed of light, the mean result required is that for the CO_2 laser f_{R12} (§ 6), uncorrected for the offset from transition centre. This offset does not contribute to the overall uncertainty, except indirectly through related effects under the heading of reproducibility in table 4. The total systematic uncertainty in the laser frequency becomes 4.0 parts in 10^{10}, which combined with the random uncertainty as above gives a total uncertainty for the CO_2 laser frequency of 4.2 parts in 10^{10}.

It should be remarked that the reproducibility of the stabilized CO_2 laser encompasses the net effect on frequency of the expected variation of running parameters for the three lasers of the same design that were used for the series of frequency and wavelength-determining experiments, over a period of about a year. Further experience of the laser behaviour, over about another year, has been used in assigning the reproducibility.

8. Discussion of frequency measurement

This experiment has the feature that, although it requires the close organization of a large number of complex pieces of equipment, its theoretical basis, namely that passive nonlinear devices generate integer multiples of incident frequencies, is exceedingly simple and hardly possible to doubt. In the case of the Josephson junction, there is the slight *a priori* possibility of internally-generated frequencies leading to incorrect intermediate frequencies (see §3a,) but for the m.i.m. diode, assuming that it is merely a passive, positive nonlinear resistor, no such possibility exists. Almost certainly then, any major sources of error which we may have overlooked will either lie among defects in the ancillary electronic equipment, or be caused by some inaccuracy in our handling of the new measurement data, e.g. by the assignment of an incorrect sideband.

To our knowledge, only one other accurate measurement has been made of a CO_2 laser frequency, that by Evenson *et al.* (1972, 1973) at N.B.S., Boulder, who determined the frequency of the adjacent R(10) transition of CO_2. Their choice of this transition largely determined our decision to measure the R(12) one, because we did not want to be influenced by their result. At N.B.S. also, Petersen and co-workers (1973) were at the time making accurate measurements of the frequencies between adjacent CO_2 lines including the R(12)–R(10) difference. However, although the N.B.S. R(12) result so obtained was available to us, though not published, shortly before we had completed our measurements, we deliberately kept it in the sealed envelope in which it was sent (K. M. Evenson: private communication), until our results were calculated.

The N.B.S. value for the R(12) frequency was 32 176 079 489 kHz, with a stabilized laser frequency uncertainty (70%-confidence) of ± 14 kHz and a transition frequency (in the gas cell) uncertainty of ± 24 kHz, containing a 20 kHz allowance for possible, unmeasured, shifts. The difference between their stabilized-laser frequency and our value for f'_{R12} is 7 kHz, 2.2 parts in 10^{10}, and between their transition frequency and our value for f_{R12}, 17 kHz. Both differences lie well within the combined 70%-confidence uncertainties so that the comparison affords convincing evidence of the reliability of both sets of measurements.

Our result for f'_{R12} with a provisional uncertainty of 30 kHz was taken to the meeting of the Comité Consultatif pour la Définition du Mètre concerned with recommending a value for c (Rowley 1973) to support Evenson's R(10) measurement and the methane-stabilized He–Ne laser frequency measurement that depended on it. Also we have since used our result to measure the same methane-stabilized laser frequency (Blaney *et al.* 1975, 1976) and this measurement agrees well with that of the N.B.S. group.

Appendix. HCN-laser asymmetry tests

A test was made to see whether counting the HCN laser frequency (as in the experiment) agreed with the result obtained by observing a beat on a spectrum analyser – the method used to transfer measurement to the H_2O laser. There is

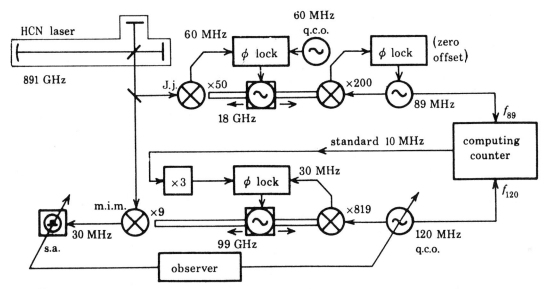

FIGURE 8. Test for asymmetric fluctuation of the HCN-laser frequency. At the top is the usual Josephson-junction 'divider' chain used for time averaging; at the bottom the spectrum-analyser beat chain, with conventional multipliers.

a possibility of a difference arising from, say, an asymmetrical time variation of the laser frequency, as remarked on in §5. The spectrum-analyser beat was observed against a klystron harmonic generated in an m.i.m. diode so that this also checked the exactness of harmonic generation in the Josephson junction (§3a). The klystron used with the m.i.m. diode was phase locked to a 120 MHz quartz crystal oscillator run from a battery (Knight, to be published), so that its frequency fluctuations did not contribute significantly to the width or potential asymmetry of the beat against the HCN laser.

The scheme of the apparatus is shown in figure 8. The usual time-averaged counting chain (§3a and 3e) is at the top and beneath it is the 'test' chain. In this a 99 GHz klystron was phase locked by a single-loop technique (Knight 1971) to a harmonic of the 120 MHz quartz-crystal oscillator so that its 9th harmonic mixed with the laser to give the beat (at 30 MHz) observed on the spectrum analyser. The beat frequency could be finely adjusted by changing the running voltage of the 120 MHz crystal oscillator. The r.m.s. frequency deviation of the crystal oscillator for 1 s averaging was about 1.3 parts in 10^{10}, about five times less than that of the HCN laser, and the multiplied phase noise 'plinth' which it and the klystron contributed to the observed beat (Knight 1971) was more than 20 dB below the carrier. Typical parameters of the beat spectrum may be compared with those given in §5 for the HCN/H_2O beat, the beat here being narrower mainly because of the lower frequency. These were: signal to noise ratio 20 dB (displayed 10 dB/div.); width about 10 kHz; spectrum analyser bandwidth, dispersion and scan rate 1 kHz, 20 kHz/div. and 20 ms/div., and timebase free-running.

To make a measurement, the operator first centred the beat on the (precalibrated) spectrum-analyser display by tuning the 120 MHz frequency, then pushed the button to start the program controlling the computing counter. This first made the time-averaged measurement, taking 1 s, and then read the 120 MHz frequency set by the operator. The space-averaged value of HCN frequency was subtracted from the time-averaged value (apart from a 'fixed' net i.f. difference) and printed, together with the actual HCN frequency (to measure the drift). The time delay in making the counts made it necessary to correct for thermal drift of the HCN laser, but the fractional drift of the 120 MHz oscillator was found to be ten times smaller and was neglected. Any possible 'settling' of the 120 MHz frequency between setting and counting was also neglected.

Two runs were made on different days, varying the observer, each run consisting of about 60 measurements taken in sets of ten. In the first run the drift rate of room temperature, and thus of HCN frequency varied widely in size and direction. Hence it was possible to plot the observed frequency difference against the HCN drift to obtain a slope corresponding to a mean delay between the two measurements. The result was 0.7 ± 0.3 s, consistent with a 0.5 s mean delay for the time average. The thermal drift was subtracted from individual results by the use of this value for the delay, and the observed drift rate. The mean correction was about 1.5 parts in 10^{10}, which was of the same order as the statistical uncertainty of each day's corrected results when combined.

The mean corrected result was:

$$f(\text{time}) - f(\text{space}) = -0.06 \pm 0.10 \text{ kHz}$$

The total uncertainty was predominantly statistical and amounted to 1.2 parts in 10^{10} (70 % confidence level). Within this limit there was no difference observed between the methods.

The current fluctuations of the laser discharge were variable, depending on the stability of the striations, and showed no noticeable asymmetry. With a current of 0.85 A the principal fluctuations were typically about 10 mA peak to peak, with components between 100 Hz and 1 MHz. The 3-phase mains ripple frequency was usually hidden in the noise. With frequency pulling of about 1 kHz/mA (Fuller, Hines & Compton 1969) the current fluctuations accounted for much of the spectral width, but their examination was a less sensitive and less complete test for asymmetry than that described above.

The tests were made about a year after the CO_2 R(12) frequency measurements, with a power supply that was different but of the same type, so that ± 2 parts in 10^{10} have been allowed for possible asymmetry in that work.

REFERENCES

Allan, D. W. 1966 *Proc. IEEE.* **54**, 221–230.
Bay, Z., Luther, G. G. & White, J. A. 1972 *Phys. Rev. Lett.* **29**, 189–191.
Bergstrand, E. 1951 *Ark. Fys.* **3**, 479.
Blaney, T. G. 1971 *J. Phys. E. Scient. Instrum.* **4**, 945–948.
Blaney, T. G., Bradley, C. C., Edwards, G. J., Jolliffe, B. W., Knight, D. J. E., Rowley, W. R. C., Shotton, K. C. & Woods, P. T. 1974 *Nature, Lond.* **251**, 46.
Blaney, T. G., Bradley, C. C., Edwards, G. J., Jolliffe, B. W., Knight, D. J. E. & Woods, P. T. 1975 *Nature, Lond.* **254**, 584–585.
Blaney, T. G., Bradley, C. C., Edwards, G. J. & Knight, D. J. E. 1971 *Phys. Lett.* **36A**, 285–286.
Blaney, T. G., Bradley, C. C., Edwards, G. J. & Knight, D. J. E. 1973a *Phys. Lett.* **43A**, 471–472.
Blaney, T. G., Bradley, C. C., Edwards, G. J., Knight, D. J. E., Woods, P. T. & Jolliffe, B. W. 1973b *Nature, Lond.* **244**, 505.
Blaney, T. G., Edwards, G. J., Jolliffe, B. W., Knight, D. J. E. & Woods, P. T. 1976 *J. Phys. D: Appl. Phys.* **9**, 1323–1330.
Blaney, T. G. & Knight, D. J. E. 1973 *J. Phys. D: Appl. Phys.* **6**, 936–951.
Blaney, T. G. & Knight, D. J. E. 1974a *J. Phys. D: Appl. Phys.* **7**, 1882–1886.
Blaney, T. G. & Knight, D. J. E. 1974b *J. Phys. D: Appl. Phys.* **7**, 1887–1893.
Bradley, C. C. & Edwards, G. J. 1973 *IEEE J. Quant. Electron.* QE-9, 548–549.
Bradley, C. C., Edwards, G. J. & Knight, D. J. E. 1972c *Radio & Electron. Engineer* **42**, 321–327.
Bradley, C. C., Edwards, G. J., Knight, D. J. E., Rowley, W. R. C. & Woods, P. T. 1972a *Phys. Bull.* **23**, 15–18.
Bradley, C. C., Edwards, G. J., Knight, D. J. E., Rowley, W. R. C. & Woods, P. T. 1972b *Progress with a determination of the speed of light; introduction and parts I, II and III: Proc. 4th Int. Conf. on Atomic Masses and Fundamental Constants, Teddington, England 1971*, pp. 295–315. New York, London: Plenum Press.
Bradley, C. C. & Knight, D. J. E. 1970 *Phys. Lett.* **32A**, 59–60.
Bradley, C. C., Knight, D. J. E. & McGee, C. R. 1971 *Electron, Lett.* **7**. 381–382.
Campion, P. J., Burns, J. E. & Williams, A. 1973 *A code of practice for the detailed statement of accuracy*. London: H.M. Stationery Office.
Daneu, V., Sokoloff, D., Sanchez, A. & Javan, A. 1969 *Appl. Phys. Lett.* **15**, 398–401.
Essen, L. 1950 *Proc. R. Soc. Lond.* A **204**, 260–277.
Evenson, K. M., Wells, J. S., Matarrese, L. M. & Elwell, L. B. 1970 *Appl. Phys. Lett.* **16**, 159–162.
Evenson, K. M., Wells, J. S., Petersen, F. R., Danielson, B. L. & Day, G. W. 1973 *Appl. Phys. Lett.* **22**, 192–195.
Evenson, K. M., Wells, J. S., Petersen, F. R., Danielson, B. L., Day, G. W., Barger, R. L. & Hall, J. L. 1972 *Phys. Rev. Lett.* **29**, 1346–1349.
Freed, C. & Javan, A. 1970 *Appl. Phys. Lett.* **17**, 53–56 and 541.
Froome, K. D. 1958 *Proc. R. Soc. Lond.* A **247**, 109–122.

Froome, K. D. & Essen, L. 1969 *The velocity of light and radio waves.* London: Academic Press.
Fuller, D. W. E., Hines, J. & Compton, B. 1969 *Electron. Lett.* **5**, 448–449.
Hocker, L. O., Javan, A., Rao, D. R., Frenkel, L. & Sullivan, T. 1967 *Appl. Phys. Lett.* **10**, 147–149.
Josephson, B. D. 1962 *Phys. Lett.* **1**, 251–253.
Josephson, B. D. 1964 *Adv. Phys.* **14**, 419–451.
Knight, D. J. E. 1969 *J. Opto-electronics* **1**, 161–164.
Knight, D. J. E. 1970 *Nat. Phys. Lab. Report* QU-8.
Knight, D. J. E. 1971 *Electron. Lett.* **7**, 383–384.
Knight, D. J. E. & Brown, E. C. 1973 *Electron. Lett.* **9**, 163–164.
Matarrese, L. M. & Evenson, K. M. 1970 *Appl. Phys. Lett.* **17**, 8–10.
McDonald, D. G., Risley, A. S., Cupp, J. D. & Evenson, K. M. 1971 *Appl. Phys. Lett.* **18**, 162–164.
Petersen, F. R., McDonald, D. G., Cupp, J. D. & Danielson, B. L. 1973 *Phys. Rev. Lett.* **31**, 573–576.
Rowley, W. R. C. 1973 *Proces Verbaux, Com. Int. Poids Mesures* **41**, 98–115.
Shotton, K. C. & Rowley, W. R. C. 1975 *Nat. Phys. Lab. Report* Qu-28.
Woods, P. T. & Jolliffe, B. W. 1976 *J. Phys. E: Sci. Instr.* **9**, 395–402.

Proc. R. Soc. Lond. A. 355, 89–114 (1977)
Printed in Great Britain

Measurement of the speed of light
II. Wavelength measurements and conclusion

By T. G. Blaney, C. C. Bradley, G. J. Edwards, B. W. Jolliffe, D. J. E. Knight, W. R. C. Rowley, K. C. Shotton and P. T. Woods

National Physical Laboratory, Teddington, Middlesex, U.K.

(*Communicated by J. Dyson, F.R.S. – Received 6 October* 1976)

Wavelength measurements have been carried out on the radiation at 9.3 μm from a stabilized CO_2 laser. The wavelength in vacuum was determined as 9 317 246 348 femtometres (fm), with a standard error of the mean of 11 fm (1.14 parts in 10^9) and a total systematic uncertainty of ± 13 fm (± 1.4 parts in 10^9). An upconversion technique of optical mixing in a non-linear crystal was used, so that the interferometric wavelength measurement was carried out on visible difference-frequency radiation at 679 nm. Its wavelength was determined relative to that of an iodine-stabilized He–Ne laser at 633 nm, by using plane-parallel Fabry–Perot interferometers.

When combined with the result of the frequency measurement described in part I, this wavelength measurement leads to a value for the speed of light in vacuum of $299\,792\,459.0 \pm 0.6$ m/s (Blaney *et al.* 1974). This result is in agreement (within the quoted uncertainties) with the value recommended internationally by the 14th General Conference of Weights and Measures in 1975.

1. Introduction

This is the second part of a paper describing a new measurement of the speed of light carried out at N.P.L. The method is based upon the measurement of both the frequency and wavelength of a stabilized laser. In part I the frequency measurement was described and in this part we describe the measurement of the wavelength. Outlines of the system that it was proposed to use for this wavelength measurement were given by Rowley & Woods (1972) and Bradley *et al.* (1972). The techniques involved are described in detail below, together with the results which, as discussed below, are now considered to have slightly smaller uncertainties than when announced briefly by Jolliffe *et al.* (1974).

The fundamental standard of length, against which the CO_2 laser wavelength has been measured, is defined in terms of the wavelength of ^{86}Kr radiation at 605 nm. In principle, the CO_2 wavelength could thus have been determined by a direct wavelength comparison using interferometric techniques, but as it lies in the infrared at 9.3 μm, this method presented a number of practical difficulties. These result partly from the large diffraction effects present at this wavelength in

[89]

any optical system of convenient dimensions, and partly from the lack of suitable optical materials for an instrument capable of operating at both visible and infrared wavelengths. Differences in the refractive indices for these two different radiations would introduce further difficulties.

These problems were avoided by using an upconversion technique for the laser wavelength measurement, as proposed by Bradley et al. (1972). In this method, the infrared radiation is superimposed upon light from a 633 nm He-Ne laser in a cooled crystal of proustite. The non-linear optical properties of this material generate radiation with a frequency exactly equal to the difference between the frequencies of the input beams, corresponding to a wavelength of 679 nm in the deep red part of the visible spectrum. Hence

$$f_{679} = f_{633} - f_{9.3}$$

so that, assuming the speed of light to be independent of wavelength (see §6)

$$1/\lambda_{679} = 1/\lambda_{633} - 1/\lambda_{9.3}$$

and $\lambda_{9.3} = \lambda_{633}/(1-R)$ where $R = \lambda_{633}/\lambda_{679}$.

In this way the wavelength of the infrared CO_2 laser is calculated from two wavelength comparisons involving only visible radiations, the first to determine λ_{633} by comparison with the ^{86}Kr standard, and the second to determine R. A measurement of λ_{633} had previously been carried out at the N.P.L. by Rowley & Wallard (1973), in the form of an interferometric comparison between a He-Ne laser stabilized to a hyperfine transition in $^{127}I_2$, and the ^{86}Kr primary standard. The pressure-scanning technique and apparatus described in §4, which had been developed over many years for accurate wavelength measurement, were used for this measurement. The uncertainty in the result was 4 parts in 10^{10}, which is much smaller than the uncertainties associated with the measurements of $(1-R)$ described below. Moreover the reproducibility of the iodine-stabilized He-Ne laser is better than 2 parts in 10^{10} over indefinite periods and between different laboratories, so that it could be regarded as a wavelength standard for this present series of measurements.

The second wavelength comparison, that necessary to determine R, is described in detail below. It was carried out with a plane parallel Fabry–Perot interferometer system, in which the mirrors are interchangeable between a long and short etalon so as to eliminate the end effect arising from dispersion of the phase change at reflexion. Two quite separate experimental techniques were employed. Initial measurements were carried out with the interferometer in a pressure-scanned mode of operation. Subsequently a more accurate servocontrol technique was developed and applied to this experiment.

It should be noted at this stage that although the upconversion technique offers considerable practical advantage over a direct wavelength measurement, it suffers from one major disadvantage. The fractional uncertainty in the measured value of R is transformed into a significantly greater uncertainty in $\lambda_{9.3}$ because of the factor $(1-R)$ in the wavelength calculation. This increases the uncertainty by a factor of fourteen – the ratio of $\lambda_{9.3}$ to λ_{679}.

In §2, the upconversion technique for generating the 679 nm difference-frequency is outlined, and in §3 the basic optical system of the interferometer is described. The two modes of operation are described in §4 and §5. Finally in §6 the results of the work described in parts I and II of this work are brought together and a new value for the speed of light deduced.

2. Upconversion

(a) *The principle and practical arrangement*

Upconversion is a technique of infrared optical mixing in a nonlinear crystal. It was first applied to the detection of weak infrared signals. The sum or difference-frequency radiation, capable of detection by photomultiplier or photodiode, was produced by mixing the weak infrared radiation with that from a powerful visible or near-infrared pump laser (Warner 1968; Gandrud & Boyd 1969; Boyd, Bridges & Burkhardt 1968). A simplified relationship between the input radiation powers P_h and P_c and the difference-frequency power P_d may be derived (e.g. Yariv 1968),

$$P_d = \frac{\omega_d^2}{\epsilon_0 c n_d n_c n_h} \chi P_h P_c L^2 \frac{\sin^2 \frac{1}{2}|\Delta k|L}{\frac{1}{2}|\Delta k|L}$$

where the suffixes h, c and d refer to the pump, infrared and difference frequencies respectively; and ω is the angular frequency; c, the speed of light *in vacuo*; n, refractive index of the crystal; χ, the nonlinear coefficient; L, length of the mixing crystal; $\mathbf{k}$, wave propagation vector; and $\Delta\mathbf{k} = \mathbf{k}_h - \mathbf{k}_d - \mathbf{k}_c$. In the equation, which assumes plane waves, the last factor, representing the well known phase-matching condition (Yariv 1968), has a maximum value of unity when $\Delta\mathbf{k} = 0$ and decreases to zero when $\frac{1}{2}|\Delta k|L = \pi$. For significant quantities of radiation to be generated, the propagation velocities within the crystal must be adjusted so that the wave vectors satisfy the condition $\mathbf{k}_d = \mathbf{k}_h - \mathbf{k}_c$.

To generate visible difference-frequency radiation from that of the 9.3 μm CO_2 laser and the 633 nm He–Ne laser, the arrangement shown in figure 1 was used. Radiation from the powerful CO_2 laser was mixed with the light from a weaker visible He–Ne laser, to give difference-frequency radiation of wavelength 679 nm. The radiations were mixed in a single crystal of proustite, cooled with liquid nitrogen. Proustite (Ag_3AsS_3) is a negative uniaxial crystal of point group $3m$, and is available as large, good quality, synthetic crystals (Bardsley *et al.* 1969).

Approximately half the radiation from the single-frequency CO_2 laser was necessary for illuminating the external gas cell used for frequency stabilization. The remainder was focused into the crystal with a barium fluoride lens, and was chopped at 400 Hz for ease of monitoring the upconverted power. Radiation from the He–Ne laser was focused separately into the crystal, at a small angle to the

CO_2 radiation. This both avoided the use of a beam splitter, and enabled tangential phase-matching to be achieved, giving a larger angular acceptance cone for the input radiation (Warner 1969). The crystal was placed in a mount fitted with fine angular adjustments so that the phase-matching condition could be optimized experimentally. The He-Ne laser, which had a discharge tube 800 mm in length, produced approximately 10 mW of single-frequency radiation, tunable over the

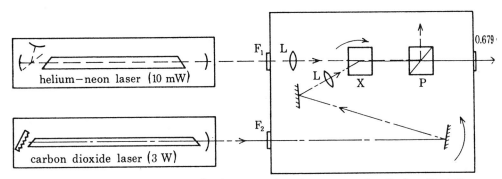

FIGURE 1. Experimental arrangement for the generation of upconverted radiation: F_1, 0.63 μm band-pass filter; F_2, infrared pass filter, P, Glan Taylor polarizer; X, proustite crystal; L, focusing lens.

bandwidth of the 633 nm transition. A mode-selecting cavity with a piezoelectric length control was used to give single-frequency operation, as described by Smith (1965). The 679 nm radiation, having its plane of polarization orthogonal to that of the 633 nm radiation, was separated from the other transmitted light by a Glan–Taylor polarizing prism. Three filters (Barr & Stroud Ltd, Type RG 645) further attenuated the residual 633 nm radiation to less than 10^{-11} W. With the crystal used for the wavelength measurements, and near-optimum focusing, 10 μW of difference-frequency radiation could be obtained from 10 W of CO_2 and 20 mW of He-Ne laser radiation. Under normal operating conditions, approximately 0.3 μW of difference-frequency radiation was obtained from 1 W of CO_2 power and 7 mW of He-Ne radiation.

(b) Optimization of output intensity

Two types of phase-matching (Midwinter & Warner 1965) are possible:

	He-Ne	CO_2	diff.
type I	e-ray	+ o-ray	→ o-ray
type II	e-ray	+ e-ray	→ o-ray

Type II conditions were used for two reasons:

(i) For neither type I or II are the phase-matching angles 90°, so that double refraction causes the ordinary and extraordinary beams to separate within the crystal. When both the incident beams are of the same polarization, however, they are more nearly collinear and a larger region of interaction results.

(ii) The absorption coefficients of proustite, shown in figure 2, indicate that there is less absorption at room temperature if the CO_2 radiation is propagated as an extraordinary ray.

An experimental investigation of the optimum operating conditions was undertaken in which the CO_2 radiation was 'chopped' with a duty cycle of 1:10,

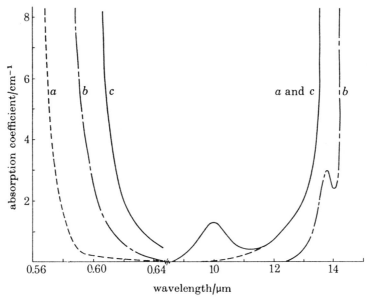

FIGURE 2. Wavelength dependence of the optical absorption coefficient of proustite: curve a, e-ray at liquid nitrogen temperature; curve b, e-ray at room temperature; curve c, o-ray at room temperature.

to avoid excessive heating and mechanical damage to the crystal. The crystal used for this investigation had a slightly different efficiency to that used later for the wavelength measurement. Figure 3 shows the dependence of the difference-frequency power output on the incident CO_2 radiation power, measured for constant He–Ne power and focusing conditions. With the room-temperature crystal the output intensity was found to saturate, an effect caused by temperature-induced absorption of the helium–neon radiation. No saturation behaviour occurred, however, when the crystal was cooled with liquid nitrogen (curve a of figure 3), and the decreased absorption coefficients at this temperature are shown in figure 2.

The effects of input-beam focusing were also investigated with the same experimental system. Both beams were chosen to have the same confocal parameter (Kogelnik & Li 1966), so that, provided they were focused at the same point, the wavefronts had common curvatures throughout the crystal. Thus although the two beam radii were different due to the unequal wavelengths, the difference-frequency power contributions from all regions of the crystal interfered constructively.

The dependence of the output intensity on this common confocal parameter is shown in figure 4 where the results were fitted theoretically (Boyd & Kleinman 1968), making allowance for the type II phase-matching. The results confirmed that, as expected for the type II arrangement, there was no measurable separation of the two input beams induced by double-refraction.

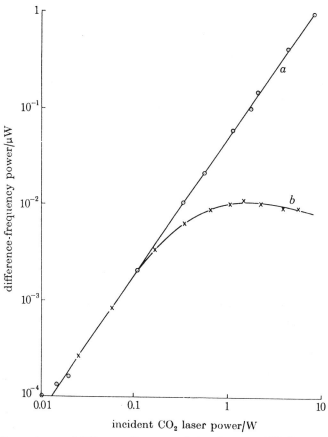

FIGURE 3. Dependence of difference-frequency intensity upon CO_2 laser power for 12 mW of He-Ne laser radiation, a confocal parameter of 108 mm and a 10 mm crystal length: curve a, liquid nitrogen temperature; curve b, room temperature.

3. DETAILS OF THE OPTICAL SYSTEM

(a) General arrangement

The interferometer and associated optical system used to measure the wavelength ratio R are shown in figure 5. The upconverted radiation and the light from a He-Ne reference laser acting as wavelength standard illuminated the Fabry-Perot etalon through separate optical channels. Each channel contained a rotating diffuser to destroy the spatial coherence of the radiations within the instrument,

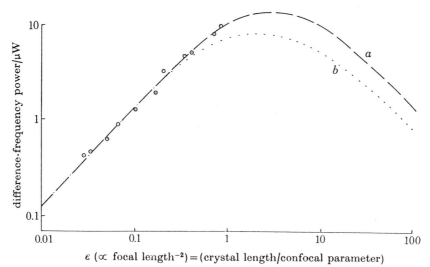

Figure 4. Variation of difference-frequency intensity with input-radiation focusing parameter and a comparison with a theoretical model. The experimental results for He–Ne and CO_2 laser powers of 12 mW and 10 W respectively are shown as discrete points. The theoretical curves were computed with two different values for the double refraction angle ρ: curve a, $\rho = 0$ rad; curve b, $\rho = 0.045$ rad.

a collimator to provide approximately parallel light at the etalon, and a diffraction grating (1800 lines/mm) to define the wavelength accepted by the instrument. The input beams were combined at the surface of a semireflector, the reflexion/transmission characteristics of which had been chosen in favour of the comparatively weak upconverted beam. The interference patterns formed by the light in passing through the Fabry–Perot etalon were focused by a lens ($f = 440$ mm) onto a pinhole exit aperture (0.3 mm in diameter), which allowed only the light from the central region to reach the photomultiplier.

(b) The Fabry–Perot etalons

A set of Fabry–Perot etalons giving reflector separations from 15 mm to 200 mm was developed for this experiment. The exploded view (figure 6) shows details of their construction. The central spacer consisted of a fused silica tube of 50 mm diameter and 3 mm wall thickness, cut with the end surfaces approximately parallel. Piezoelectric elements were attached at 120° intervals around the ends with epoxy resin cement. Each element consisted of two plates (Vernitron PZT 5H, 5 mm × 8 mm and 2 mm thick), sawn from the same sheet. These plates were cemented together in a staggered configuration which allowed electrical contacts to be soldered directly to each surface. An aluminium alloy ring was then cemented to the outer surfaces of the elements. Small invar studs cemented into blind holes within the rings, in line with each element, provided the contact surfaces for the etalon plates. The studs were then ground and polished coplanar, with the two sets parallel to about one visible fringe. Removable aluminium alloy rings served to hold the etalon plates against the invar studs. The piezoelectric pairs at one end of the spacer were used individually to adjust the parallelism of the unit, those at the other end being connected in parallel to provide length control.

The two interferometer plates used with the spacers were fused silica disks of 50 mm diameter (20 mm thickness), with a 3 mrad wedge angle, although apertures were used in the optical system to restrict the useful diameter to 25 mm. The principal surfaces were matched in flatness to $\lambda/100$ and coated with silver films of 90% reflectivity. A subsidiary pair of parallel plates was used initially with

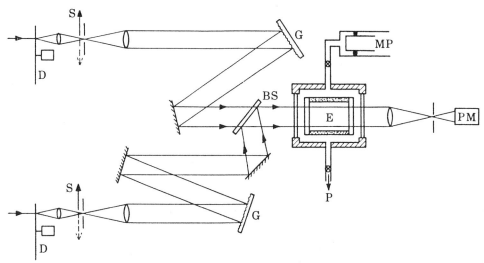

FIGURE 5. Experimental arrangement of the Fabry–Perot etalon and associated optics: D, diffusing screen; S, shutter; G, grating; BS, beam splitter, E, Fabry–Perot etalon; P, vacuum pump; MP, motorized piston; PM, photomultiplier.

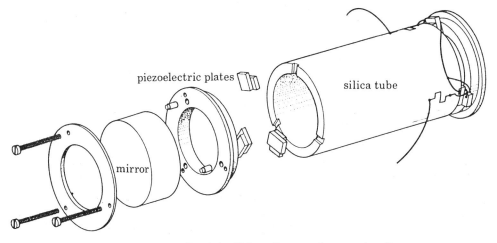

FIGURE 6. Details of the Fabry–Perot etalon construction.

each spacer to allow a measurement of its length to be made by comparison with an end bar standard gauge in a moving carriage laser interferometer (Bennett 1971). The accuracy of this technique, combined with direct interferometric measurements of the lengths of the shorter etalons with ^{86}Kr radiations, provided a knowledge of the lengths of the longer etalons to within one visible fringe.

The parallelism of the etalon was adjusted periodically throughout each series of measurements. For this purpose the interferometer was servo-controlled to a stabilized laser so that the etalon length was maintained at a value giving maximum transmitted flux through the exit pinhole. This maximum transmission is very sensitive to the average parallelism of the plates. Systematic adjustment of the voltages on two of the piezoelectric elements so as to maximize the signal, provided a simple and objective setting of etalon parallelism. This method has the useful property that the area of the plates used for adjustment is the same as that used during a wavelength intercomparison.

(c) *The etalon vessel*

The etalons were contained within a thick-walled tubular steel vessel having demountable end flanges fitted with wedged windows of fused silica. Electrical connections to the piezoelectric elements were made through vacuum 'feed-through' sockets, and the internal air temperature was measured with a thermocouple so that the refractive index could be calculated. Temperature stability was achieved by holding the vessel a few degrees above ambient with an external heater winding in a feedback control system employing a thermistor bridge. When used in the pressure scanned mode, the vessel was connected to a cylinder fitted with a piston capable of motion at a uniform speed, the piston being driven by a leadscrew and synchronous electric motor. In the servo-controlled mode of operation, the vessel was isolated and maintained at a pressure of less than 10^{-2} Pa by a sorption pump.

(d) *Etalon illumination*

Two factors were of particular importance for the correct illumination of the Fabry–Perot etalon. First, in the optical system employed with the interferometer, an image of the entrance aperture was formed in the plane of the exit pinhole. The interference fringes formed by the etalon were also focused in this plane and, in effect, modulated the spatial intensity distribution of the radiation forming that image. If the shape of the observed interference fringes was not to be distorted significantly by variations in their illumination, it was necessary to provide a uniform intensity distribution at the entrance aperture of the instrument. In practice, of course, this only applies to the area corresponding to that of the exit pinhole.

Secondly, the slight flatness imperfections of the etalon plates lead to variations in etalon length at different positions within their aperture. The final interference pattern is the summation of the patterns from all the elemental areas. Any difference in the way in which the two radiations illuminated the plates would thus have weighted this averaging process differently, and resulted in a different apparent etalon length for each radiation.

These two illumination conditions cannot be achieved directly with the coherent beam of radiation from a laser source, but require a spatially incoherent and extended source of radiation. A diffusing screen placed in each beam was used to provide such a source. With the screen imaged into the plane of the entrance aperture, the intensity distribution at the exit pinhole corresponded to the cross-sectional profile of the laser beam, and the plate illumination to that of the angular scattering characteristics of the screen.

To achieve good illumination uniformity at the exit aperture from the reference He-Ne laser, the size of the laser beam image was chosen to be considerably greater than that of the pinhole. In the channel carrying upconverted radiation, however, additional optical components were necessary. The area of illumination on the diffusing screen in this case was rectangular, extending approximately six times further in one direction than the other, and containing straight line fringes of high visibility formed in the proustite crystal. A long-focus cylindrical lens was used to equalize the aspect ratio of the beam at the diffuser, so as to utilize the available light more effectively. The systematic effect of the fringes was randomized by the changes in their positions within the beam which occurred every few minutes when the angle of the proustite crystal was readjusted for optimum 679 nm output. Additionally, for the servolock measurements, an effectively uniform image was achieved by small-amplitude angular scanning of the beam about two orthogonal axes. Sawtooth audio-frequency scans were used, having a displacement amplitude at the diffuser comparable to the beam diameter.

The illuminated areas of the two diffusers were imaged onto the entrance apertures of their collimators with a magnification of two. Glass diffusing screens, etched under a variety of conditions, were investigated for their angular scattering efficiency. The final choice of diffuser, however, was a translucent soft plastic sheet material (0.15 mm thickness) which, although having only about half the forward efficiency of a single glass screen of comparable angular scattering characteristic, gave a more uniform distribution of light.

(e) Optical adjustments

The interference pattern, in the plane of the exit pinhole, has circular symmetry with the centre corresponding to light passing normally through the etalon. Light passing through the etalon at a slight angle α has a shorter optical path difference (by a factor $\cos \alpha$) and corresponds to an annulus in the plane of the exit pinhole. Thus with a pinhole of finite size, the effective path difference of the etalon is an integration of the optical path over the angular size of the exit pinhole. The path difference is wavelength independent, so that the size and position of the pinhole are not critically important. The size of the pinhole, however, should correspond to a path difference range of only a small fraction of a wavelength, and preferably be roughly equivalent to the width of the interference maxima.

Similarly the pinhole need not in principle be centred accurately on the interference pattern; although in practice if this is not so the interference maxima observed photoelectrically are asymmetrical. Errors may then occur in relating patterns of different radiations, which will in general have different widths. To centre the pinhole in this instrument, the pinhole was mounted on the axis of a cylinder so that it could be removed and a crosswire graticule mounted in its place. An optical system, positioned in the place of the photomultiplier, illuminated the graticule through a semireflector so that the graticule could be viewed with an eyepiece. In this way the image of the graticule formed by reflexion from the Fabry–Perot etalon could be superimposed on the graticule itself, and the etalon axis adjusted to achieve good superposition. Similarly the accuracies of centring the graticule and the pinhole in their cylindrical mounts could be verified by this method, by making observations with the mounts in various rotational positions.

To achieve proper illumination of the etalon and the exit pinhole simultaneously, the area of diffuser illuminated must be on the optical axis through the centre of the etalon, with the central ray of the diffused light along this axis. This optical axis passes through the centre of the exit pinhole when it is in correct adjustment as above and through a corresponding, but larger, entrance aperture which is conjugate with, and centred with respect to, the exit pinhole. The beams of light from the two sources were aligned onto this axis by removing the diffusers and the lenses which focused them onto the entrance apertures. The beams were then adjusted to pass centrally through the entrance aperture and simultaneously through the centre of the working aperture of the etalon. For the laser wavelength standard the beam was sufficiently intense and well defined for this to be carried out by using paper screens to locate the beam position. The upconverted radiation, however, was insufficiently intense to be seen on a screen unless focused. Its beam was aligned by inserting a mirror into the optical system so that the light coming through the entrance aperture could be viewed directly. The centring with respect to the etalon aperture was also judged in transmission, by removing the photomultiplier and looking at the etalon through the exit pinhole, made larger than usual for this procedure. The input beam, which was almost parallel, was brought to a focus by the collimator lens at a position roughly coincident with the centre of the etalon. Thus when correctly aligned, a small spot of light was seen at the centre of the etalon aperture. After aligning the beams in this way, the diffusers and focusing lenses were replaced so that the diffused images could be centred with respect to the interference fringes. This adjustment was carried out by viewing the interference pattern visually with a removable auxillary optical system. Final small adjustments of the image positions were then made photoelectrically to maximize the transmitted intensity when the etalon was adjusted to the peak of an interference fringe.

4. Wavelength measurement by pressure scanning

Wavelength measurements have been carried out at the N.P.L. over a period of more than ten years by using a Fabry–Perot interferometer in a pressure-scanned mode of operation with photoelectric detection. The apparatus has been continually improved and yields results in excellent agreement with those of other laboratories (Cook 1962; Mielenz et al. 1968; Rowley 1973a). The method of operation of this equipment has been described in some detail by Rowley & Wallard (1973). For the upconversion measurements, a 633 nm Lamb-dip stabilized He–Ne laser (Spectra-Physics model 119) served as temporary wavelength standard. Its wavelength was determined, by beat frequency measurement, with respect to an iodine-stabilized laser (component 'h' of the $^{127}I_2$ spectrum at 633 nm) immediately before and after the series of wavelength measurements. The 10-mW 633 nm laser used to generate the upconverted radiation was stabilized with respect to the Lamb-dip laser with a digital offset-lock system. The Lamb-dip laser illuminated one of the two diffusers of the apparatus, as shown in figure 5, the light being chopped at 325 Hz by a rotating shutter disk immediately adjacent to the collimator entrance aperture. The 679 nm upconverted radiation illuminated the other optical channel, and was chopped at 475 Hz. Two phase-sensitive detector systems, synchronized to the chopping frequencies, amplified the photomultiplier signals so that the intensities resulting from the two radiations could be recorded digitally for subsequent computer analysis.

To make a measurement, the pressure of the dry air in the etalon enclosure was altered with the motorized piston, so that the refractive index changed steadily, thus scanning the optical path difference. Three interference maxima were recorded from each source over about 10 min, and measurements were always made in pairs, one with the pressure increasing and the other with it decreasing, so as to eliminate (and quantify) the systematic error due to inequality of the response times of the amplifying systems. The intensities were recorded at equal increments of piston movement. The slight nonlinearity of the associated refractive index increment is taken into account analytically as part of the calculation procedure, based on measurements of the pressure and temperature at the start and end of each scan. The relative positions of the interference maxima were determined from the intensity readings on both sides of each peak, at six intensity levels between 30 and 80 % of the peak height.

Observations were made with etalons 15 and 187 mm in length, so as both to determine and eliminate the effect of the difference in phase change suffered by the two wavelengths on reflexion at the etalon plates. With the shorter etalon the results of 22 scans, over the three interference maxima of each radiation, were combined to give 11 independent observations. From these, the wavelength λ', uncorrected for phase shift, of the 679 nm radiation was determined with respect to the Lamb-dip standard. The standard error of the mean of this value was $\bar{\sigma}' = 1.6$ fm (2.4 parts in 10^9). With the longer etalon the wavelength λ'', also uncorrected for phase shift, was similarly determined from 37 scans, the standard error of the mean of this set being $\bar{\sigma}'' = 0.14$ fm (2.1 parts in 10^{10}). No differences could be detected for diffusing screens with different scattering characteristics, or when two diffusing screens were used in series with the first imaged onto the second under deliberately poor focus so as to improve the image homogeneity. In all these observations the Lamb-dip laser illumination was attenuated by various factors between 100 and 3000, by means of neutral density filters, so that the shot noise from the photocurrent of the stronger source should not seriously disturb the recording of the weaker 679 nm radiation signal. It was also arranged, by choice of the lengths of the etalons, that the maxima of the interference pattern of one source occurred at approximately the minima of the other.

The wavelength of the upconverted radiation is then given by

$$\lambda = \lambda'' + (\lambda'' - \lambda')m/(M-m),$$

where m ($= 44\,492$) is the order number for the 679 nm radiation with the short etalon, equivalent to the path difference divided by wavelength, and M ($= 553\,257$) the corresponding order number for the long etalon. Thus the upconverted wavelength was determined as

$$\lambda = 679\,129\,905.57 \text{ fm}$$

with a standard error of the mean of 0.21 fm.

The etalon phase shift θ for the 679 nm radiation relative to 633 nm as phase origin is given by

$$\theta = (\lambda'' - \lambda')Mm/\lambda(M-m)$$

which yields the value $\theta = -0.00120$ (expressed as a small discrepancy of the order number) with a standard error of the mean of 0.00011. This phase shift is in excellent agreement with a value determined at about the same time with the four ^{86}Kr radiations shown in figure 7. Each of these radiations was used in turn with the 15 mm etalon to measure the apparent wavelength of the Lamb-dip laser. The corresponding phase shifts were calculated from the differences between the apparent wavelength values and the accurate value as given by a beat frequency intercomparison with the iodine-stabilized laser. As shown in figure 7, the phase values lie on a smooth curve which is well approximated by a quadratic function passing through zero phase at the wavelength 633 nm. This curve passes precisely through the value $\theta = -0.00120$ at 679 nm, confirming the self-consistency of the upconversion wavelength measurements.

The wavelength of the 9.3 μm carbon dioxide laser radiation is given by the equation

$$\lambda_{9.3} = [\lambda_{633}/(1-R)][(f_{9.3} + 20 \times 10^6)/f_{9.3}],$$

where the second term is the correction due to the 20 MHz offset of the 10 mW laser with respect to the Lamb-dip laser of wavelength λ_{633}, and R is the wavelength ratio $\lambda_{633}/\lambda_{679}$. The wavelength of the Lamb-dip laser was determined by measuring the beat frequency (35 MHz) between this laser and a laser stabilized to component 'h' of $^{127}I_2$. Using the wavelength value for the latter of 632 991 369.6 fm, as determined by Rowley & Wallard (1973), we obtained the value $\lambda_{633} = 632\,991\,416.5$ fm.

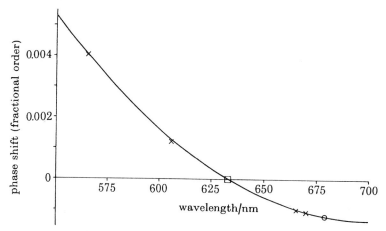

FIGURE 7. Phase shift θ (expressed as a small discrepancy of the order number). Four experimental points correspond to the ^{86}Kr radiations 565, 606, 665 and 670 nm. The two other marked points are the phase origin at 633 nm and the value $\theta = -0.00120$ determined with upconverted radiation at 679 nm.

Thus the infrared wavelength is calculated to be

$$\lambda_{9.3} = 9\,317\,246.33 \text{ pm}.$$

The uncertainty associated with this result, the contributions to which are summarized in table 1, has a statistically determined component with a standard error of the mean of 4.2 parts in 10^9 (0.04 pm) and a systematic component which is estimated at ± 2.6 parts in 10^9 (± 0.02 pm). The systematic contributions have been estimated at roughly a 70 % confidence level, and it is reasonable in this case to add in quadrature the systematic and random uncertainties, to give a combined uncertainty of ± 5 parts in 10^9 (± 0.05 pm), again to be regarded as a 70 % confidence level.

In addition to the observational statistics mentioned above for the long and short etalons, a statistical contribution is included for the wavelength measurement of λ_{633} – the relation of the iodine-stabilized laser wavelength to the primary standard – corresponding to the uncertainty quoted by Rowley & Wallard (1973). The additional uncertainty due to the use of the Lamb-dip laser is included in the list of systematic contributions.

The contribution due to refractive index effect arises from the dispersion of the air, at approximately atmospheric pressure, within the etalon. The ratio of the refractive indices for the two wavelengths under intercomparison occurs in the wavelength calculation. This ratio, which differs from unity by 5 parts in 10^7, is determined from temperature and pressure readings taken at the start and end of each scan.

Any systematic difference in the uniformity of illumination by the two radiations, in conjunction with the inevitable errors of flatness of interferometer plates, can result in systematic measurement errors. In this instrument the errors have

TABLE 1. UNCERTAINTIES OF PRESSURE SCANNING WAVELENGTH MEASUREMENT (PARTS IN 10^9).

observational statistics	$\bar{\sigma}$	estimated systematic uncertainties	
determination of λ_{633} ($^{127}I_2$) relative to ^{86}Kr	0.4	laser reproducibilities	
		Lamb-dip laser (± 1 MHz)	± 2
		carbon dioxide laser	± 0.2
determination of $(1-R)$		refractive index calculation	
with short etalon	2.8	pressure (± 0.1 torr)	± 1
with long etalon	3.1	temperature (± 20 mK)	± 0.5
		illumination non-uniformity	± 1
		chromatic aberration	± 0.7
resultant standard error of mean	4.2	quadrature summation	± 2.6

customarily been estimated, and partially corrected, by interchanging the optical channels used for the standard and test radiations. In the present series of measurements, however, this practice was unsatisfactory because of the low intensity of the upconverted beam, necessitating the use of an unequal reflexion/transmission ratio at the beam-combining semireflector. The magnitude of the effect was therefore estimated by using the standard laser radiation in place of the upconverted radiation, and measuring the apparent departure of the wavelength ratio from unity. The uncertainty contribution given for this effect represents 50% of the correction applied.

The contribution for chromatic aberration is the estimated effect due to the small wedge-angles of the final Fabry–Perot plate and of the enclosure exit window, which give rise to a wavelength dependent displacement of the Fabry–Perot ring pattern.

5. Wavelength measurement under servocontrol

(a) Electronic control system

In the servocontrol method of operation, the vessel enclosing the Fabry–Perot etalon was evacuated, and the etalon length adjusted piezoelectrically. A sinusoidal length modulation of peak to peak amplitude $\lambda/160$ at 716 Hz was applied continuously and the associated intensity modulation detected by the photomultiplier was synchronously rectified. The rectified signal is zero when the pinhole is centred on an etalon transmission peak. For small offsets from the maximum of the transmission peak, the rectified signal has a positive or negative value approximately proportional to the offset, with the sign dependent upon whether the etalon path difference is greater or smaller than that corresponding to the maximum. This signal, after passing through an integrating filter, was applied to the etalon piezoelectric element so as to form a feedback loop, keeping the etalon under servolock to a transmission maximum. The optical path difference was then an integral number of wavelengths of the radiation being used.

The electronic system used for the servolock has been described by Shotton & Rowley (1975). The amplified signals from the photomultiplier pass through a coherent filter before being rectified by a phase-sensitive detector. Both the coherent filter and phase-sensitive detector use field effect transistors as switching elements, which operate from square-wave reference signals synchronized to the oscillation of the etalon length. A dc amplifier of adjustable gain following the phase-sensitive detector provides the input for the integrating filter. This filter makes the overall gain frequency dependent, with a 6 dB per octave decrease of gain with frequency from very low frequencies to well beyond the open-loop unity gain point. A high-voltage transistor stage is then used to give an output of up to 1 kV for the piezoelectric elements.

For the determination of the wavelength ratio $R = \lambda_{633}/\lambda_{679}$, the arrangement shown in figure 8 was used. The wavelength standard λ_{633}, was a He–Ne laser stabilized to the 'd' component of $^{127}I_2$. The etalon length was maintained by the servolock at a constant value giving peak transmission of this radiation, but the feedback loop was interrupted periodically for the upconverted radiation to be passed through the interferometer. The etalon length remained constant during this interval because the electrical input to its integrating filter was interrupted, so that the filter maintained its steady voltage output.

The upconverted radiation, generated as described in §2, corresponded to the difference-frequency between the radiations from the CO_2 laser and from the 633 nm He–Ne laser of nominal 10 mW output. Thus the wavelength of the upconverted light could be tuned slightly by tuning the 10-mW laser. This tuning was carried out with a piezoelectric control on the length of the laser cavity. In this way it was possible, by a feedback control similar to that used for the etalon length, to tune the upconverted radiation for peak transmission through the Fabry–Perot etalon. Under these conditions the etalon optical path difference corresponded to integral numbers of wavelengths for both the radiations. Their wavelength ratio was equal, except for small corrections, to the ratio of these two integers. To refer the wavelength of the upconverted light to the iodine-stabilized laser standard, however, it was also necessary to measure the small wavelength difference between the 10 mW and standard lasers. This was carried out by determining their beat frequency. The timing diagram for the sequential operation of the optical shutters, amplifiers, servosystems, and beat-frequency counting is given in figure 9.

A slow response servolock (unity-gain frequency = 0.2 Hz) was used to control the etalon length in order to preserve the inherently good short-term stability of the etalon itself. With a wide bandwidth servolock system the etalon length

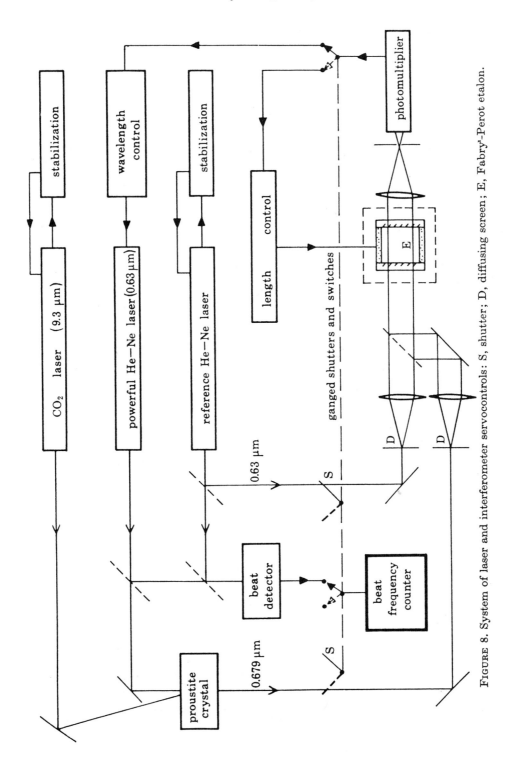

FIGURE 8. System of laser and interferometer servocontrols: S, shutter; D, diffusing screen; E, Fabry–Perot etalon.

would have been forced to follow closely the short-term fluctuations in the wavelength of the standard laser. For the control of the upconverted wavelength, however, a faster response (unity-gain frequency = 2 Hz) was appropriate because the 10 mW laser could change its frequency randomly by up to 20 MHz in the intervals when it was not under active control, and the control was required

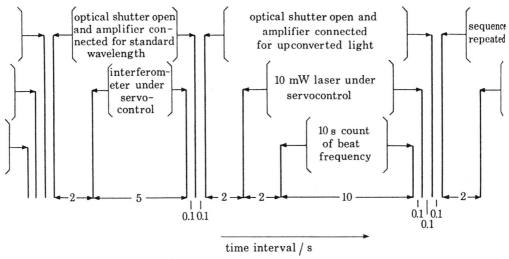

Figure 9. Timing sequence of shutter, amplifier and servocontrol operation.

to correct this within 2 s. The weak intensity of upconverted radiation available, combined with the low transmissivity of the interferometer and associated optics, resulted in significant shot noise in the photocurrent used for the servocontrol of this laser. Frequency fluctuations of about 2 MHz amplitude were thus impressed on the laser, but were averaged adequately by the 10 s count time used for the beat-frequency measurement.

During initial experiments it was found that the dominant source of noise in this servolock system was dark current from the photomultiplier (EMI type 9558B). By fixing an annular permanent magnet to the outer surface of the photocathode, however, the dark current was reduced to a level at which it did not contribute significantly to the scatter observed in our final results. No focusing optics were used after the exit pinhole of the interferometer, and the diverging beam illuminated only the central 4 mm of the photomultiplier aperture.

(b) *Illumination non-uniformity and the 'zero-beat' technique*

An important source of systematic error in the instrument arises from any differences in the illumination distribution over the aperture of the etalon for the two radiations. As discussed in §3d, errors of flatness and parallelism can combine with such illumination differences to give a systematic error. In the optical arrangement (figure 5), small illumination differences arose from the use of separate optical channels between the diffusers and combining semireflector. The magnitude of the effect was determined in the following manner.

Light from the 10 mW laser was arranged to illuminate the diffuser and optical channel normally used for the upconverted light. The servosystems should then, in the absence of systematic effects, control the etalon and 10 mW laser so that the beat-frequency averages to zero. To measure this beat frequency, the bidirectional counting technique described by Rowley (1975) was used. This is able to distinguish between positive and negative frequencies from the sequence of positive and negative amplitudes of two signals in phase-quadrature, derived from the optical phase shift at a semireflector. The beat frequency was found to be systematically shifted from zero, typically by 200 kHz with the 187 mm etalon, corresponding to a path difference discrepancy of 4 parts in 10^{10}.

The measured discrepancy was treated as a correction to be applied to the optical frequency of the standard laser radiation. This is equivalent to using the optical channel with the standard merely to maintain an arbitrary but reproducible etalon path difference, and gives the result which would be obtained if both the standard and upconverted radiations used the same optical channel. The value of the correction changed slightly after readjustment of the optics between the diffusers and the etalon and was measured during each session.

This technique of 'zero-beats', with two similar lasers illuminating the interferometer input channels, was also used to investigate other effects which could have led to systematic errors in the wavelength comparison. Thus it was shown that variations in the relative intensities of the two input beams did not significantly affect the result. Similarly the introduction of an optical chopper in one of the channels, such as that used in the CO_2 laser and hence effectively in the upconverted beam, did not cause any systematic change. Finally the small-amplitude scan imposed on the angular position of the mirrors before the diffuser in the channel to be used for upconverted light was investigated and shown to have no detrimental effect.

In all, about four thousand preliminary beat-frequency or 'zero-beat' observations were made over a period of several months. These served to optimize the control systems and to establish the conditions under which the measurements were insensitive to optical and electronic variations, and least subject to systematic bias due to the experimental conditions.

(c) Experimental procedure

When it was clear that the apparatus and technique were operating in a consistent manner, the final data were recorded over a series of daily sessions. The experimental procedure adopted was dictated by the period for which various parts of the apparatus remained within acceptable limits of adjustment. At the start of each day the 10 mW laser was adjusted for maximum single-frequency power, the output of upconverted radiation optimized and the beat-frequency system adjusted for maximum signal. The axes of the two input beams were then aligned with the interferometer and the positions of all optical components subsequently placed in these beams were monitored to ensure no resultant displacement of the axes. Electronic offset voltages were set for zero integrator drift with the servocontrol circuits complete but no signal falling on the photomultiplier. Finally the etalon parallelism, autocollimation and the positions of the images of the diffusing screens at the entrance apertures were adjusted, following the procedures described in § 3.

A measurement block extending over 5 min, and comprising ten individual 10 s counts of the beat frequency was possible without disturbing the apparatus in any way. At that stage, however, it was necessary to refill the liquid nitrogen dewars of the CO_2-laser fluorescence detector and the proustite crystal and, because of the sensitivity of the phase-matching angle, to re-optimize the crystal position for maximum 679 nm output. After three such blocks, occupying a period of 30–40 min, it was occasionally found that the etalon was out of parallelism by more than the $\lambda/100$ adjustment reproducibility. Hence after each group of three blocks the etalon parallelism, autocollimation and entrance-aperture image positions were checked.

After three or four groups of results the electronic offsets were reset and the rate of etalon length drift measured by recording a series of counts of the beat frequency with the 10 mW laser locked to the etalon, but without the usual servocontrol of etalon length between each count. Provision was also made to allow rapid exchange of the upconverted radiation in one channel for attenuated radiation direct from the 10 mW laser. It was thus possible to alternate groups of measurements between those necessary to obtain the wavelength ratio R, and the 'zero-beat' measurements necessary to update the channel difference correction.

The result of each 10 s count was recorded on paper tape to facilitate computer analysis of statistical variations and the application of drift rate and other corrections. A number of such day-long sessions were devoted to measurements with each of two etalons (employing the same mirrors) which had different lengths. Combination of the results with both etalons enables the effect of the wavelength-dependent phase change of light to be eliminated, as described below.

(d) Calculation of the wavelength

The equation quoted in §1 relating the frequency of the upconverted radiation to those of the 9.3 μm laser and the 633 nm stabilized laser wavelength standard, must be modified so as to take into account the beat-frequency B between the 10 mW laser and the standard. Thus the 10 mW laser has a frequency $(f_{633}+B)$, so that
$$f_{679} = f_{633} + B - f_{9.3}.$$
This leads to the equation
$$\lambda_{9.3} = [\lambda_{633}/(1-R)] [1-(B/f_{9.3})].$$

The wavelength ratio R is now given by the ratio of the integral numbers M_{679} and M_{633} of the upconverted and standard radiation wavelengths corresponding to the

etalon path difference, with a correction for the optical phase shift θ_{679}, relative to the standard
$$R = (M_{679} - \theta_{679})/M_{633}.$$

The phase term can be eliminated by combining observations made with etalons of different lengths. Thus observations were made with the long etalon (plate separation 187 mm) for which the integral order number for the standard was $M_{633} = 593\,426$, and for the upconverted light $M_{679} = 553\,110$; and the short etalon (plate separation 19 mm) for which the corresponding numbers were $m_{633} = 60\,835$, and $m_{679} = 56\,702$.

Thus the complete equation for calculating the infrared wavelength is
$$\lambda_{9.3} = \lambda_{633}[\Delta M_{633} - (M_{633}\,B_2/f_{9.3}) + (m_{633}\,B_1/f_{9.3})]/[\Delta M_{633} - \Delta M_{679}]$$
where $\Delta M_{633} = M_{633} - m_{633}$, $\Delta M_{679} = M_{679} - m_{679}$, and B_1, B_2 are the beat-frequencies observed for the short and long etalons respectively.

An experimental difficulty occurs when a short etalon is operated in the servo-lock mode, because the beat frequency can be inconveniently high. The beat frequency is given by
$$B = f_{9.3}[1 - \lambda_{9.3}(1-R)/\lambda_{633}],$$
where R is the ratio of order numbers given above. Thus it may be calculated that, with the 19 mm etalon, changing the optical length by one order number (for both radiations) changes the beat frequency by 529 MHz. The beat frequency initially did not lie, for the piezoelectric tuning range of $\Delta m = 4$, within the ± 150 MHz range of the available frequency-counting system. It was necessary to grind down the etalon spacer to a slightly shorter length. Changing the value of m_{633} by 15, and m_{679} by 14, altered B by 148 MHz. Thus from a rough measurement of the frequency with a spectrum analyser, it was possible to calculate the change of spacer length necessary to bring the beat within the range of the counter.

(e) *Wavelength measurement results*

The observations used to calculate the final wavelength value extended over a period of several weeks. During this time six measurement sessions, comprising 900 observations (30 blocks) of the beat frequency were carried out with the 187 mm etalon, and two sessions (11 groups) with the 19 mm etalon. The variations between the individual observations within a block, between blocks and between groups were all entirely self-consistent. The correction for the systematic illumination effect was determined for each session from two groups of 'zero-beat' measurements at the beginning and two at the end of a session. The scatter of the corrected results for the different sessions was statistically consistent with the scatter within each session. This concordance of the results was further confirmation that adjustment of the apparatus and changes of alignment due to temperature changes did not systematically affect the measurements.

The value for B_1 obtained by combining the results with the short etalon was -124.725 MHz with a standard error of the mean of 0.237 MHz. With the long etalon, B_2 was measured as -57.325 MHz with a standard error of the mean of 0.019 MHz. Thus the mean corrected result for the wavelength of the CO_2 laser was determined as

$$\lambda_{9.3} = 9\,317\,246\,348 \text{ femtometres.}$$

This result has an uncertainty due partly to (statistically determined) random effects, and partly arising from (estimated) systematic effects. The statistical uncertainty, corresponding to a standard error of the mean $\bar{\sigma} = 11$ fm (1.14 parts in 10^9), is given by the quadrature summation of the contributions from the

TABLE 2. UNCERTAINTIES OF SERVOLOCK WAVELENGTH MEASUREMENT (PARTS IN 10^9)

observational statistics	$\bar{\sigma}$	estimated systematic uncertainties	
determination of λ_{633} ($^{127}I_2$) relative to ^{86}Kr	0.4	reproducibility of $\lambda_{9.3}$	± 0.2
		reproducibility of λ_{633}	± 0.2
		drift of long etalon	± 0.3
determination of $(1-R)$		drift of short etalon	± 0.6
with long etalon	0.66	illumination non-uniformity	± 0.7
with short etalon	0.84	chromatic aberration	± 0.7
		intensity modulation of λ_{633}	± 0.7
resultant standard error of mean	1.14	quadrature summation	± 1.4

determinations of B_1 and B_2, together with the statistical uncertainty of the wavelength measurement of λ_{633} by Rowley & Wallard (1973). The contributions to the estimate of the systematic uncertainty are listed in table 2, in which the total systematic uncertainty is derived by quadrature summation as ± 13 fm (± 1.4 parts in 10^9). If the random and systematic uncertainties are added together in quadrature, as for the pressure scanning measurement, the combined uncertainty of the 9.3 μm wavelength measurement by the servocontrol method amounts to ± 17 fm (± 1.8 parts in 10^9). This uncertainty is much smaller than that determined by the pressure-scanning method (see §4), and both results are in excellent agreement, so that this servocontrol result can be regarded as the final result of the wavelength measurement. Its uncertainty is smaller than that previously announced (Jolliffe *et al.* 1974), as a simple addition of the random and systematic contributions was previously used, whereas it is now considered that a quadrature summation is more appropriate.

The systematic uncertainty contribution in respect of the etalon length drift arises because of incomplete compensation for this effect by the 'zero-beat' observations. Although the effect of drift is in principle eliminated in this way, an uncertainty arises because the drift rate was not always constant during a session. Furthermore the intervals of a few seconds between the end of the etalon servocontrol and the centre of the beat-frequency count were not identical for the two types of observation. The effects of variations in the illumination uniformity and their determination by the 'zero-beat' method have already been discussed in some detail. The uncertainty remaining from this source is assessed as 10% of the correction. The uncertainty included for the intensity modulation of the $^{127}I_2$ laser arises because the laser was stabilized by a third harmonic method which allows a residual intensity modulation corresponding to the fundamental of the frequency modulation. The time-averaged centre of the modulation will thus be dependent upon whether the averaging system is or is not sensitive to intensity. Thus the wavelength centre, as detected interferometrically, will differ slightly from the frequency centre determined by a beat-frequency observation. This effect was investigated experimentally on component 'f' of $^{127}I_2$, for which the fundamental intensity modulation was ten times as great as with component 'd'.

6. THE SPEED OF LIGHT: DISCUSSION AND CONCLUSION

The experiments described above and in part I, yield values for the frequency (Blaney *et al.* 1973) and free-space vacuum wavelength (Jolliffe *et al.* 1974) of the radiation from a CO_2 laser stabilized to the R(12) transition of CO_2 at 9.3 μm. By multiplying these values together, one obtains a value for the speed of light.

With the frequency and wavelength values obtained in these experiments ($f_{9.3}$ = 32 176 079 482 kHz and $\lambda_{9.3}$ = 9 317 246 348 fm), the value (Blaney *et al.* 1974) for the speed of light is

$$c = 299\ 792\ 459.0 \pm 0.6\ \text{m/s}.$$

The overall uncertainty (equivalent to $\pm 2.0 \times 10^{-9}$) quoted here is the quadrature sum of the standard error of the mean (1.1×10^{-9}) and the total systematic uncertainty ($\pm 1.6 \times 10^{-9}$). Each of these was derived by quadrature summation of the contributions from both the frequency and wavelength measurements. It must be pointed out, however, that an uncertainty value derived from such a combination of random and systematic contributions into a single number, which provides a general indication of overall uncertainty, must be treated with caution as the systematic contribution may include estimated quantities which cannot usually be based on a definite confidence level (Campion, Burns & Williams 1973). Our overall uncertainty is smaller than previously announced (Blaney *et al.* 1974) due to the use of quadrature error summations throughout, and should be considered as being roughly equivalent to a 1σ (70%) confidence level.

The frequency measurements were carried out with respect to a crystal oscillator which was stable and accurate to a few parts in 10^{10}, and which was checked periodically with a rubidium frequency standard. This standard was itself compared with the frequency of a caesium clock, so that the frequency measurements were firmly based upon the primary standard of time interval.

The wavelength measurement was carried out by reference to a 633 nm He–Ne laser stabilized to a saturated absorption feature of $^{127}I_2$, which as mentioned above is reproducible to better than 2 parts in 10^{10}. A wavelength measurement by Rowley & Wallard (1973) served to relate this stabilized laser to the ^{86}Kr primary standard. The wavelength value used for the 'd' hyperfine component was

632 991 178.3 fm, with a statistical uncertainty of 0.4 part in 10^9. The ^{86}Kr primary standard itself, however, is not perfectly reproducible, but no uncertainty contribution has been added on this account. Thus the measurements reported here relate to the 'as maintained' N.P.L. metre.

It is generally accepted (Rowley 1973a) that the lamps normally used to realize the ^{86}Kr standard radiation are reproducible to at least 1 part in 10^9, but there are pressure and Doppler-shift corrections which introduce further uncertainty. Additionally, an asymmetry of the spectral profile is frequently observed (Rowley & Hamon 1963; CCDM 1970; Barger & Hall 1973), although other measurements suggest that it is either very small or non-existent (Bayer-Helms & Engelhard 1962; Rowley 1973b; Hamon & Chartier 1973). A procedure has been suggested by Giacomo (1973) by which observations made where asymmetry was observed can be made compatible with those made where asymmetry was not detected. The difficulties in making accurate measurements with the ^{86}Kr standard have, however, been resolved such that six laboratories in different countries, with the use of different interferometric methods, have reported concordant results for the wavelength of the iodine-stabilized laser with a scatter of only ± 3 parts in 10^9. On consideration of these results and their scatter, the Comité Consultatif pour la Définition du Mètre judged that the length standard was being realized in the various laboratories with uncertainties not greater than ± 4 parts in 10^9 (Rowley 1973a). The result of Rowley & Wallard (1973), however, has a much smaller measurement uncertainty and it is, happily, in agreement with the mean of the other reported values. It is thus concluded that the wavelength assigned to the iodine-stabilized laser used in this determination of c, although based upon the N.P.L. realization of the metre, is nevertheless in excellent agreement with the realization of the length unit in other laboratories.

The value of c which we have measured may be compared directly with the value derived from a frequency and wavelength measurement by Evenson et al. (1972) of the methane-stabilized He-Ne laser at 3.39 μm. After adjustment of the wavelength value according to the subsequent study of Giacomo (1973) concerning the definition of the metre in relation to ^{86}Kr asymmetry, this was $c = 299\,792\,457.4 \pm 1.1$ m/s.

Although there is good astronomical evidence to confirm that the speed of light in free space is the same to 2 parts in 10^{15} for radiations of widely different frequency (Bay & White 1972), it is interesting to consider whether, if there were any difference, the value which we have obtained corresponds to that for the infrared radiation of the CO_2 laser, or to one of the visible radiations used in the interferometric intercomparison. If we bear in mind that the interferometric experiment determines the wavelength ratio $R = \lambda_{633}/\lambda_{679}$ and the upconversion gives a frequency difference, it may be shown that the speed of light which we have obtained is related to the values c_{633} and c_{679} at the wavelengths 633 nm and 679 nm by

$$c_{\text{measured}} = c_{679} - [c_{679} - c_{633}] \, [679/(679-633)].$$

This equation is of the same form as that relating group velocity u with phase velocity v in a dispersive medium, $u = v - \lambda \mathrm{d}v/\mathrm{d}\lambda$. Thus if the speed of light were to be slightly wavelength dependent, our measurement would correspond closely to the vacuum group velocity at 679 nm.

The experimental measurements relating to the speed of light were reviewed by the Comité Consultatif pour la Définition du Mètre in June 1973. It was concluded that the most accurate value could be derived by taking the frequency measurement of Evenson et al. (1972) referred to above, together with the wavelength measurements of the methane-stabilized laser reported by four laboratories. The value so derived was $c = 299\ 792\ 458.3$ m/s, but in view of the uncertainty in the wavelength measurements arising from the difficulty of realizing the length unit, it was recommended that the following rounded value be adopted for general use,

$$c = 299\ 792\ 458\ \text{m/s}.$$

This was considered to have an uncertainty of ± 4 parts in 10^9.

This recommended value has been endorsed by the International Committee of Weights and Measures and the 14th General Conference of Weights and Measures, and adopted as a conventional value by the International Astronomical Union. Our measurement is clearly in agreement with this value, within its uncertainty, and serves as independent confirmation that the value is in accordance with the units of length and time interval. Our work was based on a different CO_2 transition from that measured by Evenson et al. (upon which the C.C.D.M. value relied) and the wavelength determination involved visible rather than infrared laser radiation. The agreement of the speed of light values is thus a demonstration of the reliability of the techniques involved in measuring infrared frequencies and of intercomparing wavelengths.

We would like to express our thanks to the many people who helped and encouraged us throughout the course of this experiment, and in particular to K. D. Froome, A. Horsfield, N. W. B. Stone and the N.P.L. Time and Frequency group. For the provision of special experimental equipment and practical assistance, we are indebted to J. D. Dromey for his work on the lasers concerned with the frequency measurement, and to A. J. Wallard for the iodine-stabilized He-Ne laser and associated measurements.

References

Bardsley, W., Davies, P. H., Hobden, H. V., Hulme, K. F., Jones, O., Pomeroy, W. & Warner, J. 1969 *Opto-electronics* **1**, 29–31.
Barger, R. L. & Hall, J. L. 1973 *Appl. Phys. Lett.* **22**, 196–199.
Bay, Z. & White, J. A. 1972 *Phys. Rev.* **D5**, 796–798.
Bayer-Helms, F. & Engelhard, E. 1962 *Com. Consult. Definition Metre (Com. Int. Poids Mesures)* 3rd Session, 83–89.
Bennett, S. J. 1971 *Phys. Bull.* **22**, 397–398.

Blaney, T. G., Bradley, C. C., Edwards, G. J., Jolliffe, B. W., Knight, D. J. E., Rowley, W. R. C., Shotton, K. C. & Woods, P. T. 1974 *Nature, Lond.* **251**, 46.

Blaney, T. G., Bradley, C. C., Edwards, G. J., Knight, D. J. E., Woods, P. T. & Jolliffe, B. W. 1973 *Nature, Lond.* **244**, 504.

Boyd, G. D., Bridges, T. J. & Burkhardt, E. G. 1968 *IEEE J. Quant. Elect.* **QE-4**, 515–519.

Boyd, G. D. & Kleinman, D. A. 1968 *J. appl. Phys.* **39**, 3597–3639.

Bradley, C. C., Edwards, G. J., Knight, D. J. E., Rowley, W. R. C. & Woods, P. T. 1972 *Phys. Bull.* **23**, 15–18.

Campion, P. J., Burns, J. E. & Williams, A. 1973 *A code of practice for the detailed statement of accuracy.* London: H. M. Stationery Office.

CCDM 1970 *Com. Consult. Definition Metre (Com. Int. Poids Mesures)* 4th Session, M29–32.

Cook, A. H. 1962 *Com. Consult. Definition Metre (Com. Int. Poids Mesures)* 3rd Session, 129–148.

Evenson, K. M., Wells, J. S., Petersen, F. R., Danielson, B. L., Day, G. W., Barger, R. L. & Hall, J. L. 1972 *Phys. Rev. Lett.* **29**, 1346–1349.

Gandrud, W. B. & Boyd, G. D. 1969 *Optics Commun.* **1**, 187–190.

Giacomo, P. 1973 *Com. Consult. Definition Metre (Com. Int. Poids Mesures)* 5th Session, M126–132.

Hamon, J. & Chartier, J.-M. 1973 *Proces Verbaux, Com. Int. Poids Mesures* **41**, 35–37.

Jolliffe, B. W., Rowley, W. R. C., Shotton, K. C., Wallard, A. J. & Woods, P. T. 1974 *Nature, Lond* **251**, 46–47.

Kogelnik, H. & Li, T. 1966 *Appl. Opt.* **5**, 1550–1567.

Midwinter, J. E. & Warner, J. 1965 *Br. J. appl. Phys.* **16**, 1135–1142.

Mielenz, K. D., Rowley, W. R. C., Wilson, D. C. & Engelhard, E. 1968 *Appl. Opt.* **7**, 289–293.

Rowley, W. R. C. 1973a *Proces Verbaux, Com. Int. Poids Mesures* **41**, 98–115.

Rowley, W. R. C. 1973b *Com. Consult. Definition Metre (Com. Int. Poids Mesures)* 5th Session, 94–99.

Rowley, W. R. C. 1975 *J. Phys. E: Scient. Instrum.* **8**, 223–226.

Rowley, W. R. C. & Hamon, J. 1963 *Rev. Opt.* **42**, 519–531.

Rowley, W. R. C. & Wallard, A. J. 1973 *J. Phys. E: Scient. Instrum.* **6**, 647–652.

Rowley, W. R. C. & Woods, P. T. 1972 *Proceedings of the 4th conference on Atomic Masses and Fundamental Constants, held at Teddington England, Sept 1971* (edited by Sanders J. H. & Wapstra, A. H.) pp. 311–315. London: Plenum Press.

Shotton, K. C. & Rowley, W. R. C. 1975 *Nat. Phys. Lab. Report* Qu-28.

Smith, P. W. 1965 *IEEE J. Quant. Elect.* **QE-1**, 343–348.

Warner, J. 1968 *Appl. Phys. Lett.* **12**, 222–224.

Warner, J. 1969 *Opto-electronics* **1**, 25–28.

Yariv, A. 1968 *Quantum electronics.* New York: Wiley & Sons.

Precise Frequency Measurements in Submillimeter and Infrared Region

YURI S. DOMNIN, NICKOLAY B. KOSHELJAEVSKY, VICTOR M. TATARENKOV, AND PAVEL S. SHUMJATSKY

Abstract—As a result of frequency stabilization of all the lasers and microwave oscillators in a new multiplication chain consisting of HCN (337 μm), D_2O (84 μm), CO_2/OsO_4 (10.53 μm), CO_2 (10.18 μm), and He–Ne/CH_4 (3.39 μm) lasers signals up to 88 THz have been synthesized with high precision. Owing to phaselocking of HCN and D_2O lasers to the primary frequency standard synthesis accuracy of $\sim 10^{13}$ up to 30 THz has been achieved for the first time. The frequency of the CO_2/OsO_4 laser was first measured

$$\nu_{OsO_4} = 28\,464\,676\,938.5 \pm 1 \text{ kHz}$$

and the He–Ne/CH_4 laser frequency was determined to be

$$\nu_{CH_4} = 88\,376\,181\,586 \pm 10 \text{ kHz}.$$

TO SOLVE THE problem of absolute laser frequency measurement we have proposed and realized step-by-step the following new chain for frequency conversion from a cesium primary standard to the infrared (IR) region [1]:

Cs standard $\xrightarrow{1}$ HCN (337 μm) $\xrightarrow{2}$ D_2O (84 μm) $\xrightarrow{3}$

$\xrightarrow{3}$ CO_2 (10.53 μm) $\xrightarrow{4}$ CO_2 (10.18 μm)

$\xrightarrow{5}$ He–Ne (3.39 μm).

Here we indicated the operating substances of the lasers—their wavelength and number of conversion steps. The He–Ne laser was frequency stabilized to the nonlinear absorption peak in methane [3]; in the same way the CO_2 (10.53 μm) laser was frequency stabilized using absorption in $^{192}Os^{16}O_4$ vapor [2]. Besides the lasers, microwave sources in the centimeter or millimeter region are necessary at each step of the chain. Frequency conversion was produced by special point-contact mixer diodes, W–Si for the first step and W–Ni at others.

Manuscript received June 28, 1980.
The authors are with the State Service of Time and Frequency, Gosstandart, Moscow, U.S.S.R.

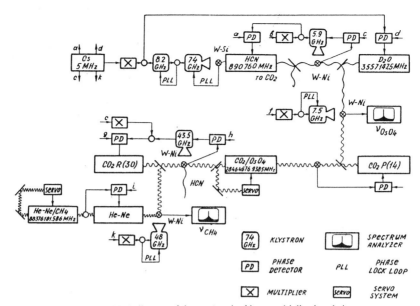

Fig. 1. Block diagram of the synchronized laser multiplication chain.

Due to the frequency stabilization and synchronization of all lasers and microwave oscillators we have substantially increased the accuracy of absolute frequency measurements in the submillimeter (SM) and IR regions. For the first time we have synthesized signals up to 30 THz with the accuracy of the primary standard. Using these signals we have measured for the first time the CO_2/OsO_4 absolute laser frequency, and have determined the He–Ne/CH_4 laser frequency [6] with five times greater accuracy than previously [4], [5].

Two factors contributed decisively to the increased accuracy. The first factor is a rational design of multiplication chain, that is, incorporating new and promising intermediate frequency lasers and standards in comparison with [4], [5]. The use of D_2O and CO_2/OsO_4 lasers enabled a decrease in the order of multiplication and enabled satisfactory signal-to-noise ratios to be obtained at each step—not less than 20 dB in a bandwidth of 50 kHz. Secondly, we succeeded in synchronizing both of the SM lasers to the frequency standard using phaselock techniques. Here we see the main sources of success. The block diagram of the locked chain is shown in Fig. 1. Fig. 2(a), (b) gives general front and rear views of the installation.

The SM laser frequency was stabilized to the cesium primary standard using a special phaselock system for the D_2O laser. The unique feature lies in its very narrow bandwidth. The idea of such a phaselock system was first put forward and realized using an HCN laser [7]. The concept is that with relatively high short-term laser stability combined with long-term stability of the standard the bandwidth of the phaselock loop may ultimately be very narrow, of the order of few hertz. In this case such a phaselock system will act as narrow-band tracking filter suppressing the noise of the multiplied signal from the standard. The filtering properties of such systems are not inferior to a superconducting cavity and considerably surpass it in accuracy. The narrow-band phaselock system of the D_2O laser is realized in practice in the following way. The HCN laser, 74-GHz klystron, and 8.2 GHz klystron are successively synchronized to the D_2O laser in a wide band. The signal from the standard at 8.2 GHz is compared in phase with the klystron radiation and an error is supplied to the mirror drive of the D_2O laser via the narrow-band filter. A slow D_2O laser frequency drift is almost completely tracked and does not affect the measurement accuracy for averaging times greater than 10 s. When the system is in lock all the above oscillators acquire well-defined frequency values. Particularly, ν_{HCN} = 890 760 MHz and ν_{D_2O} = 3 557 147.5 MHz with the accuracy of the primary standard. The measurement of laser frequencies in the SM and IR regions including the CO_2 laser range with an accuracy of $\sim 10^{13}$ became possible using HCN and D_2O lasers themselves and their harmonics. Using the 8th harmonic of the D_2O laser we can now determine the frequency of the CO_2/OsO_4 standard with the accuracy of the primary standard.

The CO_2/OsO_4 laser frequency with an external absorption cell is stabilized to the most powerful saturation peak in OsO_4 in the CO_2 P(14) laser region. Unlike [3] we tried to form a pure standing wave in the absorption cell. For this purpose the absorption was low and a fully reflecting mirror was placed behind the cell. The bandwidth was ~ 200 kHz and contrast ~ 2 percent. According to our estimate the laser frequency stability was $\sim 3 \cdot 10^{-11}$ during the period of a day. The measurements of the CO_2/OsO_4 laser frequency were performed during a month at possibly more constant parameters of the laser and servosystem. The most difficult problem in this experiment was to control the frequency offset in the CO_2/OsO_4 laser due to spurious amplitude modulation. At the beginning of each series of measurements the offset was carefully compensated and at the end any existing unbalance was recorded in order to make a correction. As a result the following value for the CO_2/OsO_4 laser frequency has been obtained:

$$\nu_{OsO_4} = 28\ 464\ 678\ 938.5 \pm 1 \text{ kHz}.$$

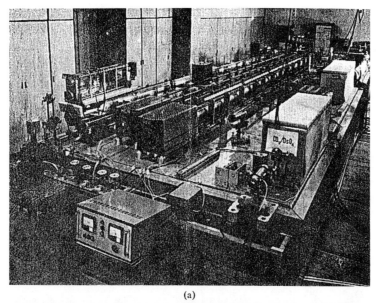

Fig. 2. (a) General front view of the experimental installation (photograph). (b) General rear view of the experimental installation (photograph).

The uncertainty is almost completely determined by the CO_2 laser instability.

Knowing now the frequency of the CO_2/OsO_4 laser and using steps 4 and 5 of the multiplication chain we have measured the He–Ne/CH_4 laser frequency. In this case the CO_2 $R(30)$ laser frequency was synchronized to the CO_2/OsO_4 laser frequency by a special phaselock system. The principle of its operation is shown in Fig. 1. The beat frequency of the 3rd harmonic of the CO_2 $R(30)$ laser, a powerful ($\sim$100 mW) He–Ne laser phaselocked to the He–Ne/CH_4 standard and 48-GHz klystron was measured by spectrum analyzer. The He–Ne/CH_4 laser has the following characteristics. It is a portable instrument with a sealed discharge and absorption cell. The saturation peak full bandwidth $\sim$300 kHz and peak amplitude $\sim$3 percent of output power. The resettability of the laser frequency is better than $1 \cdot 10^{-11}$, long term stability $\sim 10^{-12}$ for several hours. After the data processing of several series of measurements we have obtained the value of the He–Ne/CH_4 laser frequency:

$$\nu_{CH_4} = 88\ 376\ 181\ 586 \pm 10 \text{ kHz}.$$

The 10-kHz estimated uncertainty is due mainly to the CO_2/OsO_4 laser frequency instability. In that experiment MOM diode back reflection strongly affected the frequency of the CO_2/OsO_4 standard. This effect was reduced by a Fresnel rhomb and a small misalignment, but we did not succeed in excluding it fully and it contributed 3-kHz uncertainty to the CO_2/OsO_4 laser frequency.

Let us note that the He–Ne/CH_4 laser frequency measured in our laboratory lies within the experimental uncertainties of previously performed measurements: 88 376 181 627 ± 50 kHz [4] and 88 376 181 608 ± 43 kHz [5]. This is of great importance since the value of the methane transition frequency was used in a precise determination of the speed of light [8]. Our result has been obtained with an essentially different multiplication chain, with more stable intermediate lasers and enhances the reliability of the accepted value for the speed of light.

We note that a new stage of absolute laser frequency measurements has begun with the above work fulfillment. That stage is characterized by modern frequency standard accuracy. Frequency standards in the radio frequency region now have the possibility of being accurately compared with laser frequencies. Thanks to this the idea of a unified frequency, time and length standard may become a reality [9]. Experiments in general physics such as verification of the constancy of the more important "constants of nature" also become a real possibility [10].

Acknowledgment

We should like to express our thanks to V. S. Letokhov, O. N. Kompanets, and A. R. Kukudzanov for the CO_2/OsO_4 laser design; V. V. Nikitin, M. A. Gubin, and A. D. Tyurikov for provision of the laser servosystem; and V. S. Solov'ev, A. S. Kleiman, and I. V. Tomashko for provision of microwave sources.

References

[1] Y. S. Domnin, V. M. Tatarenkov, P. S. Shumjatsky, "CO_2 P(14) laser frequency measurement," *Sov. Phys.*, vol. QE 2(12), pp. 2612-2614, 1975.

[2] Y. A. Gorokhov, O. N. Kompanets, V. S. Letokhov, V. A. Gerasimov, and Yu. J. Posudin, "Narrow stabilization resonances in the spectrum of OsO_4 induced by CO_2 laser radiation," *Opt. Commun.*, 7, p. 320, 1973. O. N. Kompanets, A. R. Kukudzanov, V. S. Letokhov, and E. L. Michailov, "Frequency stabilized CO_2 laser using OsO_4 saturation resonances," in *Proc. Second Frequency Standard Metrol. Symp.*, July 5-7, pp. 167-186, 1976.

[3] R. L. Barger and J. L. Hall, "Pressure shift and broadening of methane line at 3.39 μm studied by laser saturated molecular absorption," *Phys. Rev. Lett.*, vol. 22(1), pp. 4-8, 1969.

[4] K. M. Evenson, J. S. Wells, F. R. Petersen, B. L. Danielson, and G. W. Day, "Accurate frequencies of molecular transitions used in laser stabilization: The 3.39-μm in CH_4 and the 9.33 and 10.18-μm transitions in CO_2," *Appl. Phys. Lett.*, vol. 22(4), pp. 192-195, 1973.

[5] T. G. Blaney, G. J. Edwards, B. W. Jolliffe, D. J. E. Knight, and P. T. Woods, "Absolute frequency of the methane stabilized He-Ne laser (3.39) and the CO_2 R32 stabilized laser (10.18)," *J. Phys. D: Appl. Phys.*, vol. 9(9), pp. 1323-1331, 1976.

[6] Y. S. Domnin, N. B. Kosheljaevsky, V. M. Tatarenkov, and P. S. Shumjatsky, "Absolute laser frequency measurements in IR region," *ZhETF Pisma*, vol. 30(5), pp. 273-275, 1979.

[7] Y. S. Domnin, V. M. Tatarenkov, and P. S. Shumjatsky, "Laser converter of the standard frequency into the submillimeter range," *Sov. Phys.*, vol. QE 4(5), pp. 1158-1160, 1977.

[8] K. M. Evenson, F. R. Petersen, B. L. Danielson, G. W. Day, R. L. Barger, and J. L. Hall, "Speed of light from direct frequency and wavelength measurements," *Phys. Rev. Lett.*, vol. 29(19), pp. 1346-1349, 1972. T. G. Blaney, C. C. Bradley, C. J. Edvard, B. W. Jolliffe, D. E. J. Knight, W. R. C. Rowley, and K. C. Shotton, "Measurements of speed of light," in *Proc. Roy. Soc. London*, vol. A355, pp. 61-114, 1976.

[9] C. H. Townes, *Advances in Quantum Electronics*, J. S. Singer, Eds. New York; London, England: 1961, 3.

[10] Y. S. Domnin, N. B. Kosheljaevsky, V. M. Tatarenkov, P. S. Shumjatsky, O. N. Kompanets, A. R. Kukudzanov, V. S. Letokhov, and E. L. Michailov, "CO_2/OsO_4 laser—absolute frequency and new possibilities," *ZhETF Pisma*, vol. 30(5), pp. 269-272, 1979.

# Accurate Absolute Frequency Measurements on Stabilized CO₂ and He–Ne Infrared Lasers

ANDRE CLAIRON, BRAHIM DAHMANI, AND JACQUES RUTMAN, MEMBER, IEEE

Abstract—In our laboratory, we have measured the frequencies of CO_2 and He–Ne lasers near 30 and 88 THz, stabilized, respectively, by saturated fluorescence in CO_2 and saturated absorption in CH_4. Our measurement system includes a stable free-running optically pumped CH_3OH laser at 4.25 THz replacing the noisy H_2O laser used as a transfer oscillator in early experiments. As a result of the reduced mixing orders (≤ 9), beat notes between lasers are now observed with ≈ 30-dB signal-to-noise (S/N) ratios in a 100-kHz bandwidth. Therefore, beat frequencies can be measured accurately with digital counters and simultaneous counting of the frequencies involved largely eliminates the uncertainties due to transfer oscillators. The measurements are referred to the cesium beam frequency standard. The results are processed by a desktop calculator which also controls the measurement process.

INTRODUCTION

THE ADVENT OF nonlinear point contact diodes working in the submillimeter and infrared regions has extended absolute frequency measurements by harmonic mixing into these regions of the electromagnetic spectrum where lasers are used as coherent signal sources. The first laser frequency measurement was performed in 1967 on the discharge HCN gas laser near 890 GHz [1]. Successive harmonic mixing experiments have lead to successful frequency measurements of the H_2O, CO_2 and He–Ne discharge gas lasers, respectively, near 10, 30, and 88 THz [2], [3]. Of special interest for metrology are the frequency measurements performed on CO_2 and He–Ne lasers using frequency stabilization by saturated fluorescence or saturated absorption, since these devices can be considered as secondary frequency standards in the infrared. These first measurements used the HCN and H_2O discharge gas lasers as intermediate oscillators in the laser chain. Metal-insulator-metal (MIM) diodes were employed with success as nonlinear mixers up to 197 THz [4], [5], but did not extend to the measurement of the He–Ne line at 260 THz (1.15 μm). Nonlinear crystals have therefore been used to reach this frequency by addition of lower laser frequencies and also to reach the visible region by frequency doubling [6]. For frequencies lower than about 3 to 4 THz, other nonlinear devices such as point-contact Josephson junction or Schottky diodes can be used [7], [8].

In the meantime, optically pumped submillimeter lasers had been discovered and their interest for frequency metrology pointed out [9], both as a possible future standard (assuming the development of adequate frequency stabilization techniques) or as an intermediate low noise oscillator in a laser frequency measurement chain. Recently, the use of the 4.25-THz CH_3OH laser as a transfer oscillator between a 99-GHz klystron and the $CO_2R(32)$ laser, with a Josephson junction as harmonic mixer between the klystron and the CH_3OH laser, has been reported at NPL; this scheme has produced a new and precise measurement of the frequency of the He–Ne(CH_4) laser [10].

In our laboratory, we have completed new measurements of the frequency of stabilized CO_2 and He–Ne lasers with a scheme wherein the noisy H_2O discharge laser used in earlier chains has been replaced by a free-running optically pumped CH_3OH waveguide laser at 4.25 THz as intermediate oscillator between a phaselocked HCN waveguide laser[1] and the saturated fluorescence stabilized $^{13}CO_2P(28)$ laser developed in our laboratory.

A NEW LASER CHAIN INCLUDING AN OPTICALLY PUMPED (FIR) CH_3OH LASER

Time domain frequency stability measurements made in our laboratory on free-running optically pumped CH_3OH waveguide lasers working at 4.25 THz (70.5 μm) have clearly demonstrated their high-frequency stability with a two-sample standard-deviation of $\sigma_y (\tau = 50$ ms$) \approx 2 \cdot 10^{-12}$ in a 5-MHz bandwidth [11]. The frequency of the CH_3OH line at 70.5 μm can be used as an intermediate step between the HCN frequency and the frequency of some of the CO_2 laser lines. Specifically, we have used the following relationships:

$$5 \nu_{HCN} - \nu_{CH_3OH} \approx 202.129 \text{ GHz} \quad (1)$$

$$\nu_{CO_2} - 7 \nu_{CH_3OH} \approx 8.943 \text{ GHz} \quad (2)$$

where ν_i denotes the frequency of the useful laser lines: $\nu_{HCN} \approx 891$ GHz, $\nu_{CH_3OH} \approx 4.25$ THz, and $\nu_{CO_2} \approx 29.771$ THz for the $^{13}CO_2P(28)$ line.

The difference frequencies appearing in (1) and (2) can be downconverted to tens of megahertz by heterodyning, respectively, with the third harmonic of an 67.4-GHz klystron and with the fundamental of an X-band klystron the signals of which are radiated on to the diodes. Point contact W-Ni diodes are used for both harmonic mixing experiments.

The complete klystron and laser chain linking the He–Ne(CH_4) optical frequency standard to the cesium beam

---

Manuscript received June 28, 1980. This work was supported by Research Contracts from the Bureau National de Métrologie, Paris, France.

The authors are with the Laboratoire Primaire du Temps et des Fréquences, 61, avenue de l'Observatoire, Paris, France 75014.

[1] Developed by MM. Auvray, Gastaud, Pyée, and Sentz at the Paris VI University (Laboratoire d'Electronique et de Résonance Magnétique).

frequency standard has the following advantages:
- Utilization of a stable optically pumped CH_3OH laser instead of the noisy discharge H_2O laser;
- a mixing order of only 9 for the HCN-CH_3OH link (versus 14 for the HCN-H_2O link in previous chains) thus allowing higher signal-to-noise ratios;
- the line $^{13}CO_2 P(28)$ at 29.771 THz can be linked to the $^{12}CO_2 R(30)$ line at 29.442 THz with only one 65-GHz klystron whose fifth harmonic is used (whereas the third harmonic of the HCN laser was also required in earlier chains relying on $^{12}CO_2 R(30)$ and $^{12}CO_2 R(10)$ at 32.134 THz [2]).

Improvement of the various elements in the chain yield beat notes with S/N ratios higher than 20 dB thus allowing frequency measurements by counting techniques (replacing less precise determinations with spectrum analyzers as required in some experiments). As a consequence, simultaneous counting of the beat signal frequencies can be achieved, thereby canceling the effects of intermediate laser frequency instabilities in the final measurement precision.

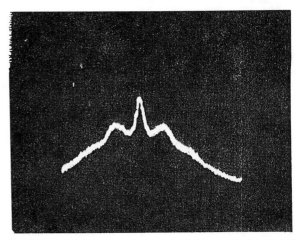

Fig. 1. Spectrum of the beat note between the 12th harmonic of the 74-GHz klystron and the HCN laser (vertical scale: 10 dB/cm; horizontal scale: 50 kHz/div., 100 ms/cm). Video filter: 100 Hz. BW: 10 kHz

DESCRIPTION OF THE LASERS

Since it has some effects on the ultimate performance of the lasers, let us recall some environmental characteristics: the experiment is carried out in an underground laboratory which is not related to any building. All the lasers and MIM diodes are located on the same 10-m-long 60-ton concrete table supported by 10 pneumatic elements.

The 3-m-long HCN laser uses a waveguide resonator comprising a 5-cm diameter Pyrex tube maintained at 150 °C to increase the efficiency and keep the tube clean [12]. Output coupling is made through a grid which fixes the polarization. The gas mixture under a total pressure of 1 torr contains 77 percent He, 16 percent CH_4, and 7 percent N_2. The discharge current is equal to ≈ 1.25 A. Output powers of about 50-60 mW are obtained with a stable discharge. The 3-dB linewidth of the free-running laser signal spectrum is about 10 kHz and is essentially due to discharge instabilities. Phaselocking the laser to a high spectral purity quartz pilot reduces its long-term frequency drift but does not decrease the linewidth in a significant way because of fast variations in the discharge current. Laser frequency control is made through its discharge current (≈ 1.8 kHz/mA) which influences the plasma refractive index. The beat note between the laser and the 12th harmonic of the 74-GHz klystron is 15-20 dB above the phase noise pedestal measured in a 10-kHz bandwidth (Fig. 1).

The optically pumped CH_3OH laser uses a 2-m-long waveguide cavity comprising a 28-mm inner diameter Pyrex tube and two gold-coated plane mirrors. Cavity length is stabilized by invar rods; one mirror is mounted on a PZT translator for fine frequency tuning. Input coupling of the pump power provided by a CO_2 laser and output coupling of the FIR 4.25-THz beam are made through the same 3-mm hole at the center of one of the mirrors. FIR and pump radiations are separated by a KCl window at Brewster angle for the CO_2 beam and which reflects about 90 percent of the 4.25-THz radiation [11]. With 20 W of pump power, 90-100 mW are now obtained at 4.25 THz in monomode operation.

The two CO_2 sealed lasers developed in our laboratory comprise 4 invar rods defining a 1.3-m-long cavity closed by a 150 grooves/mm grating (efficiency ≈ 95 percent) and a 4-m radius of curvature Zn-Se mirror with 15-percent transmission mounted on a PZT. A variable aperture iris allows monomode operation. Frequency stabilization is achieved by the saturated fluorescence technique [13] with an external cell. Feedback towards the laser is prevented by tilting the return mirror ($\approx 5 \cdot 10^{-3}$ rad) for the $^{13}CO_2 P(28)$ beam and by using a $\lambda/4$ waveplate and a grid polarizer for $^{12}CO_2 R(30)$. The fluorescence signal is detected by a large area In-Sb detector ($\phi = 7$ mm) with a cooled filter to increase the detectivity to values of $D^*_{4.3\mu m} \geq 10^{11}$ cm $\cdot$ $Hz^{1/2}$ W^{-1}.

With 2 W of laser beam power, a 12-mm diameter beam in the cell, and 30-40 mT of CO_2 pressure, inverse Lamb dips with 12-15-percent contrast and 1-1.3-MHz width are observed. After various parameter optimization, short-term stability is given approximatively by $\sigma_y(\tau) = 2 \cdot 10^{-11} \tau^{-1/2}$ for 10^{-2} s $\leq \tau \leq 10^2$ s for intense lines such as $P(20)$ and $P(18)$ of $^{12}CO_2$, i.e., 2-3 $\cdot 10^{-12}$ for $\tau = 100$ s. A weaker line such as $^{13}CO_2 P(28)$ has a stability of about 10^{-11} for $\tau \geq 10$ s. Long-term stability and reproducibility are limited by several phenomena such as synchronous detection drift, feedback towards the laser (for which no perfect solution has been found), frequency shift between the amplification line and the absorption line in low pressure CO_2 (pressure shift), distorsions and wavefront curvature in the absorption cell. Reproducibilities of about $3 \cdot 10^{-11}$ for the $^{12}CO_2 R(30)$ line, and of about 10^{-10} for the $^{13}CO_2 P(28)$ line have been measured. In some mixing experiments, the beat note linewidth can be reduced by using an unmodulated CO_2 laser which is offset frequency locked to a (modulated) fluorescence stabilized one.

The 8-m-long He-Ne transfer laser at 88 THz has its cavity closed by a plane output coupling mirror with 50-percent transmission and a 30-m radius of curvature mirror with 1-percent transmission (used for its stabilization). Three invar rods are used for length stabilization. Output power is about 60 mW with a discharge current of 30 mA in a gas mixture

^{22}Ne–He in a one to nine ratio under a pressure of 3 torr. This laser is offset locked on a metrological CH$_4$-saturated-absorption stabilized He–Ne laser[2] [14]. The 60-cm-long cavity stabilized by 3 silica rods comprises the 35-cm gain tube and the 20-cm internal absorption cell. It is closed by one multilayer dielectric mirror transmitting 4 percent at 3.39 μm and one totally reflecting mirror, with radius of curvature of 2.4 m.

HARMONIC MIXING EXPERIMENTS

A prerequisite for the realization of reliable series of precise laser frequency measurements is the generation of beat notes with S/N ratios $\geq$ 20 dB in a 100-kHz BW. As a result of many improvements and optimizations made on the lasers, MIM diodes and optical coupling of laser beams on to the diodes, S/N ratios of about 30 dB are now routinely observed in our harmonic mixing experiments. This is also due partly to the choice of the 4.25 THz CH$_3$OH laser yielding mixing orders limited to 9 (respectively, 9, 9, 7, 5 for the mixings between HCN, CH$_3$OH, ^{13}CO$_2$, ^{12}CO$_2$, and He–Ne). Frequency measurements are attempted only when S/N ratios are near 30 dB. Moreover, the use of an optically pumped laser and the good laboratory environment yield beat notes with narrower linewidths than those obtained in chains including the H$_2$O laser.

The HCN–CH$_3$OH beat note has a 3-dB linewidth of 50–60 kHz (HCN phaselocked), to be compared with 100–150 kHz for the HCN–H$_2$O experiment (mixing order: 14).

The CH$_3$OH–CO$_2$ beat note has a linewidth of 20 kHz with a stabilized CO$_2$ laser and 3 kHz with a free-running unmodulated CO$_2$ laser, versus 300–500-kHz spectral widths observed for the H$_2$O–CO$_2$ mixing stage. The CO$_2$–HeNe beat note has a 100-kHz 3-dB linewidth.

LASER FREQUENCY MEASUREMENTS

Due to the limited number of klystrons and synchronizers presently available in our laboratory, the measurement of the He–Ne(CH$_4$) laser frequency has been made in three steps (Fig. 2):
1) absolute measurement of the frequency of the saturated-fluorescence-stabilized ^{13}CO$_2$P(28) laser;
2) measurement of the frequency difference between the two CO$_2$ lines involved in our scheme (^{13}CO$_2$P(28) and ^{12}CO$_2$R(30));
3) measurement of the He–Ne(CH$_4$) frequency relying on the previous results.

Step 1) is the most difficult one since it requires simultaneous operation of 5 phase locked klystrons and 5 lasers (phaselocked HCN; free-running CH$_3$OH optically pumped by CO$_2$; unmodulated ^{13}CO$_2$ offset locked on the stabilized ^{13}CO$_2$). The HCN frequency is obtained through measurements made either on the phaselocked 9.275-GHz klystron or on the high stability (10^{-12} for $\tau = 1$–10 s; $<10^{-10}$/day) 5-MHz quartz pilot from which all the IF frequencies are synthesized. The klystron frequency measurements made over 1 s have a pre-

[2] Developed by MM. Brillet, Cérez, and Hartmann at the Paris XI University (Laboratoire de l'Horloge Atomique).

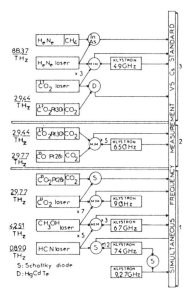

Fig. 2. Schematic representation of our laser harmonic mixing experiments (3 steps).

cision limited by the cesium standard fluctuations (model HP 5061 A). Quartz pilot or klystron frequency measurements with a 100-s averaging time allow a resolution of better than 10^{-11}. The beat signal between the HCN laser and the 12th harmonic of the 74-GHz klystron is counted in order to check the correct operation of the HCN phaselocking (lasting usually several minutes). Beat note frequencies between lasers are simultaneously counted and checked on spectrum analyzers (HP 8553B-8552B). The 67.4-GHz klystron is phaselocked to a low noise quartz oscillator-multiplier (Austron 1120 S-1125 S) via an 8.4-GHz klystron the frequency of which is counted with a microwave counter. The following counting processes are synchronized to within a few microseconds: klystron at 9.275 GHz; beat notes: HCN–CH$_3$OH, CH$_3$OH-unmodulated ^{13}CO$_2$ and unmodulated ^{13}CO$_2$-stabilized ^{13}CO$_2$.

Step 2) is much simpler since only 2 stabilized CO$_2$ lasers and 2 klystrons (65 GHz and X band) are involved; the lasers are frequency modulated ($\pm$400 kHz) with the same phase and amplitude in order to avoid beat note spectrum spreading. The 65-GHz klystron is phaselocked on the 7th harmonic of an X-band klystron, the frequency of which is counted, itself phaselocked to a quartz oscillator with high spectral purity and good long-term stability (Austron 1120 S).

In step 3), the transfer He–Ne laser is offset-locked to the metrological He–Ne(CH$_4$) laser and an unmodulated CO$_2$ laser is offset-locked on the stabilized CO$_2$ laser. The 48-GHz klystron is measured as above simultaneously with the beat note frequency and with the beat frequencies between both CO$_2$ and both He–Ne lasers. In each step, a desktop computer yields the mean, the standard deviation and the histogram of series of data.

Measurements in step 1) were taken on 4 days (April 23 and 28, 1980 and May 5 and 6, 1980) with 4 series of 10 nonadjacent data each day, some experimental changes being done between successive series. Integration time was equal to 1 s.

TABLE I

Date	Number of Readings	mean (kHz)	Standard (kHz) deviation
23 April 1980	39	29 770 668 151.6	2.2
28 April 1980	39	29 770 668 151.6	1.4
5 May 1980	38	29 770 668 151.8	2.2
6 May 1980	40	29 770 668 151.4	1.7

Only 4 results from the 160 attempted measurements have been rejected before data processing due to some obvious apparatus malfunction (delocking of klystrons or of the HCN laser). Equipment modifications were made from one day to the next such as CO_2 laser refilling, CO_2 absorption cell refilling, permutation of the CH_3OH–CO_2 beat note sidebands, and permutation of counting systems at some beat notes.

The results obtained for each day are summarized in Table I. The overall mean value is equal to 29 770 668 151.6 kHz with a standard-deviation of 1.9 kHz. Statistical tests (t and F) applied to the data indicate a certain discrepancy between series of 10 measurements. It must be pointed out that these series are not fully equivalent in the sense that their durations range from several minutes to about one hour due to some equipment failures (delocking of klystron or the HCN laser, MIM diode adjustment, optically pumped-CO_2 pump lasers adjustments). The observed discrepancy between series can be attributed to the lasers' long-term frequency instabilities. The results show that no significant differences arise from either the permutation of counting systems or of the sideband of the CH_3OH-$^{13}CO_2$ beat note. Correct operation of the counter used to measure the HCN frequency has been checked with an accuracy of $\pm 2 \cdot 10^{-12}$ (10^3-s integration time) by using two cesium beam standards. The correct harmonic relationship between the 9.275-GHz klystron frequency and the 5-MHz quartz oscillator frequency has also been checked at the $\pm 2 \cdot 10^{-12}$ accuracy level.

Measurements in step 2) were taken in two weeks. Quasi-continuous control of the beat frequency was possible due to the greater simplicity of this experiment. Integration times from 1 to 100 s were used; adjacent measurements performed over 1 and 10 s have been grouped to yield an equivalent ensemble of 151 measurements with a 100-s integration time. Roughly the same number of measurements were performed each day. From all these measurements, the difference between the two CO_2 lines is found to be

$$\bar{\nu}_{^{13}CO_2P(28)} - \bar{\nu}_{^{12}CO_2R(30)} = 328\ 184\ 836.9 \pm 3.8\ \text{kHz}\ (1\sigma).$$

During these experiments, modifications were made to the optical systems used for CO_2 laser stabilization, the CO_2 cells were refilled and a permutation of the locking electronics was made. As in step 1), statistical tests applied to the data have shown a certain discrepancy between series which can be attributed to the CO_2 laser reproducibility. The final value for our $^{12}CO_2R(30)$ laser is then

$$\bar{\nu}_{CO_2R(30)} = 29\ 442\ 483\ 314.7 \pm 4.3\ \text{kHz}\ (1\sigma).$$

Measurements in step 3) were taken in two weeks and grouped as above into an equivalent ensemble of 200 measurements with $\tau = 100$ s. The final result is

$$\bar{\nu}_{He-Ne(CH_4)} - 3\bar{\nu}_{CO_2R(30)} = 48\ 731\ 683.9 \pm 3\ \text{kHz}\ (1\sigma)$$

yielding for our He–Ne(CH_4) laser

$$\bar{\nu}_{He-Ne(CH_4)} = 88\ 376\ 181\ 628.0 \pm 13.5\ \text{kHz}\ (1\sigma).$$

The relative uncertainty is equal to about $1.6\ 10^{-10}$. Following this measurement, our He–Ne(CH_4) laser was taken to the BIPM where comparisons were performed with similar lasers previously used in international experiments (with NPL, PTB and VNIIFTRI). A frequency difference of 10 ± 2.5 kHz was measured and attributed to our laser modulation distortion. The "corrected" frequency of our laser is thus equal to 88 376 181 618.0 $\pm$ 13.8 kHz. The histograms of the data pertinent to the three steps are shown in Fig. 3(a)-(c).

Determination of the CO_2 Transition Frequencies

Corrections must be added to our CO_2 laser frequency measurements to obtain the frequencies of the CO_2 transitions.

A) The shift between the top of the CO_2 laser power tuning curve and the center of the fluorescence dip yields an offset of the frequency of the stabilized laser which depends on parameters such as the dip width and contrast. This effect has been measured as follows: adjustment of the laser frequency near the center of the dip; suppression of the reflected beam in the cell; measurement of the residual signal at the lock-in amplifiers; introduction of an offset equal to two times that signal; control of the symmetry of the derivative of the dip (after that procedure, the derivative curve appears to be symmetric to within 1 part in 300); measurement of the stabilized laser frequency. The corrections obtained are

+ 6 kHz $\pm$ 2 kHz for the $^{13}CO_2P(28)$ line
+ 7 kHz $\pm$ 2 kHz for the $CO_2R(30)$ line.

B) Beam wavefront curvature in the cell induces frequency shifts that can be determined by using the results of [15]. Our experimental arrangement and the associated uncertainties in the determination of the beam parameters give an estimate of the correction equal to ± 1.5 kHz.

C) Residual feedback into the laser can also produce an offset which is difficult to estimate precisely. Observations made with strong feedback show that it is of about ± 1 kHz when the system is well aligned.

D) Modulation distortion and integrator offset of the servo-loop can induce a ± 0.5-kHz shift on the laser frequency.

E) The pressure shift has not been measured since a constant pressure of 40 $\pm$ 5 mT has been used during all the experiments.

As a consequence, our final values for the CO_2 transition frequencies for a 40-mT pressure are

$^{13}CO_2P(28)$: 29 770 668 157.6 $\pm$ 3.4 kHz
$^{12}CO_2R(30)$: 29 442 483 321.7 $\pm$ 5.1 kHz.

Conclusion

We have measured for the first time the frequency of the $^{13}CO_2P(28)$ laser. The relative uncertainty of $\pm 6.4 \cdot 10^{-11}$ is mainly due to the reproducibility of our stabilized CO_2 laser.

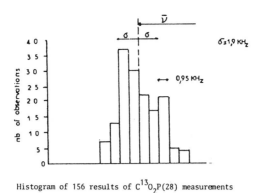
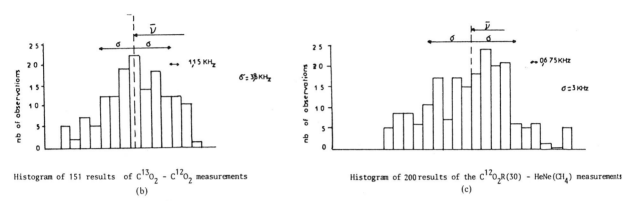

Fig. 3. Histograms of the experimental data. 3(a) step 1. 3(b) step 2. 3(c) step 3.

At this precision level, the laser chain does not contribute appreciably to the final uncertainty which is about 10 times smaller than in earlier measurements near 30 THz. Our $R(30)$ value is consistent with previous results and about 5 to 6 times more precise. Our value for the He–Ne(CH_4) laser is also consistent with all the earlier values obtained at NBS and NPL at the 6-8 10^{-10} level, and also with the very precise ($\pm 3 \cdot 10^{-11}$) recent measurement at NPL [10]. However, it is not consistent with a recent determination made in USSR [16] which is itself not consistent with the latest NPL value.

ACKNOWLEDGMENT

The authors would like to acknowledge the contributions of many people to this program during the past few years; at LPTF: A. Comeron, J. Jimenez, F. R. Petersen (NBS), J. Petersen (Copenhagen), P. Plainchamp; outside LPTF: J. Auvray, C. Gastaud, M. Pyée, A. Sentz; L. Henry; A. Brillet, P. Cérez and F. Hartmann.

Also, we would like to thank Mr. J. M. Chartier (BIPM) for his help during the He–Ne lasers intercomparisons.

REFERENCES

[1] L. Hocker et al., "Absolute frequency measurement and spectroscopy of gas laser transitions in the far infrared," Appl. Phys. Lett., vol. 10, no. 5, pp. 147-149, Mar. 1, 1967.
[2] K. M. Evenson et al., "Accurate frequencies of molecular transitions used in laser stabilization: The 3.39 μm transition in CH_4 and the 9.33 and 10.18 μm transitions in CO_2," Appl. Phys. Lett., vol. 22, no. 4, pp. 192-195, Feb. 15, 1973.
[3] T. G. Blaney et al., "Absolute frequencies of the methane stabilized HeNe laser (3.39 μm) and the $CO_2R(32)$ stabilized laser (10·17 μm)," J. Phys. D: Appl. Phys., vol. 19, pp. 1323-1330, 1976.
[4] D. Jennings et al., "Extension of absolute frequency measurements to 148 THz; Frequencies of the 2.0 and 3.5 μm Xe laser," Appl. Phys. Lett., vol. 26, no. 9, pp. 510-511, May 1, 1975.
[5] K. Evenson et al., "Laser frequency measurements: A review, limitations, extension to 197 THz (1.5 μm)," in Laser Spectroscopy III, Ed. J. L. Hall and J. L. Carlsten. New York: Springer, 1977.
[6] K. M. Baird et al., "Extension of absolute frequency measurements to the visible: Frequencies of ten hyperfine components of iodine," Opt. Lett., vol. 4, no. 9, pp. 263-264, Sept. 1979.
[7] M. Pyée and J. Auvray, "Génération et comparaison des fréquences dans le domaine sub-millimétrique et infrarouge," Bull. d'information BNM, vol. 6, no. 19, pp. 11-22, Jan. 1975.
[8] D. J. E. Knight and P. T. Woods, "Application of nonlinear devices to optical frequency measurement," J. Phys. E: Sci. Instrum. vol. 9, pp. 898-916, 1976.
[9] C. O. Weiss et al., "Optically pumped far-infrared laser for infrared frequency measurements," in Proc. AMCO 5, Ed. J. H. Sanders and A. H. Wapstra. New York: Plenum, 1976.
[10] D. J. E. Knight et al., A 3 parts in 10^{11} measurement of the frequency of the methane stabilized helium-neon laser at 88 THz," this issue, pp. 000-000.
[11] P. Plainchamp, "Frequency instability measurements of the CH_3OH optically pumped laser at 70.5 and 118 μm," IEEE J. Quant. Electron., vol. QE-15, pp. 860-864, Sept. 1979.
[12] P. Belland, "Lois d'échelle pour la conception optimale des lasers HCN (337 μm) continus utilisant l'effet de guide d'onde," thèse de doctorat d'état, Université Paris VI, Nov. 3, 1975.
[13] C. Freed and A. Javan, "Standing wave saturation resonances in the CO_2 10.6 μm transitions in a low-pressure room-temperature absorber gas," Appl. Phys. Lett., vol. 17, no. 2, pp. 53-56, July 1970.
[14] A. Brillet et al., "Frequency stabilization of the HeNe lasers by saturated absorption," IEEE J. Quant. Electron., vol. QE-10, pp. 526-528, June 1974.
[15] C. J. Bordé et al., "Saturated absorption lineshape: Calculation of the transition time broadening by a perturbation approach," Phys. Rev. A, vol. 14, no. 1, July 1976.
[16] Yu S. Domnin et al., P. S. Pis' ma v Zh. Eksp. Teor. Fiz. (USSR), vol. 30, pp. 273-275, 1979.

Improved Laser Test of the Isotropy of Space

A. Brillet[(a)] and J. L. Hall

Joint Institute for Laboratory Astrophysics, National Bureau of Standards and University of Colorado, Boulder, Colorado 80309

(Received 20 November 1978)

> Extremely sensitive readout of a stable "etalon of length" is achieved with laser frequency-locking techniques. Rotation of the entire electro-optical system maps any cosmic directional anisotropy of space into a corresponding frequency variation. We found a fractional length change $\Delta l/l = (1.5 \pm 2.5) \times 10^{-15}$, with the expected $P_2(\cos\theta)$ signature. This null result represents a 4000-fold improvement on the best previous measurement of Jaseja *et al.*

Our conventional postulate that space is isotropic represents an idealization of the null experiments of Michelson and Morley[1] and the later improved experiments of Joos.[2] Lorentz[3] and FitzGerald[3] showed that a specific longitudinal contraction could account for the null result. In his study of the axiomatic basis of the special theory of relativity, Robertson[4] has shown how this result may be combined with similarly idealized experimental results from the Kennedy-Thorndike[5] and Ives-Stilwell[6] experiments to lead unambiguously to the special theory of relativity, assuming the constancy of the speed of light. He shows that—between two inertial frames moving along x—the metric transforms as

$$ds^2 = dt^2 - c^{-2}(dx^2 + dy^2 + dz^2),$$
$$ds'^2 = (g_0 dt')^2 - c^{-2}[(g_1 dx')^2 + g_2^2(dy'^2 + dz'^2)]. \quad (1)$$

The special theory of relativity corresponds to $g_0 = g_1 = g_2 = 1$. A recent Letter summarizes the excellent agreement obtained in a wide variety of precision experiments with the predictions of special relativity.[7] Still, major advances in the three fundamental experiments are clearly of strong scientific interest, since in general we have only finite experimental limits for the velocity or absolute orientational dependences of the g_i. For example, the Joos version[2] of the Michelson-Morley experiment shows that $g_2/g_1 - 1 = (0 \pm 3) \times 10^{-11}$.

The sensitivity advantage of laser frequency metrology for length measurements was first pointed out by Javan and Townes, and with co-workers they were the first to apply laser techniques to a Michelson-Morley-type measurement.[8] Unfortunately their length etalons were in fact lasers, and a serious systematic frequency shift (275 kHz) was observed as the apparatus rotated. Thus, although their excellent intrinsic laser stability (~30 Hz) gave a glimpse of future metrological possibilities, the large spurious systematic effect limited their test for cosmic anisotropy to $g_2/g_1 - 1 = \pm 2 \times 10^{-11}$. This value is about 10^{-3} of the "predicted ether drift," based on Earth's orbital velocity $[(v/c)^2 \simeq 10^{-8}]$ and represents only a small improvement over the Joos result. The present paper extends the null result by a factor 4000 below the value of Jaseja *et al.*,[8] to a frequency shift limit of $\pm 2.5 \times 10^{-15}$, corresponding to $\pm 5 \times 10^{-15}$ in $g_2/g_1 - 1$.

Our experiment has been designed to be clear in its interpretation and free of spurious effects. Its principle may be understood by reference to Fig. 1. A He-Ne laser ($\lambda = 3.39$ μm) wavelength is servostabilized so that its radiation satisfies optical standing-wave boundary conditions in a highly stable, isolated Fabry-Perot interferometer. Because of the servo, length variations of this cavity—whether accidental or cosmic—appear as variations of the laser wavelength. They can be read out with extreme sensitivity as a frequency shift by optically heterodyning a portion of the laser power with another highly stable laser, provided in our case by a CH_4-stabilized[9] laser. To separate a potential cosmic cavity-length variation from simple drift, we arranged to rotate the direction of the cavity length by mounting the length etalon, its laser and optical accessories, onto a 95-cm × 40-cm × 12-cm granite slab which, along with servo and power-supply electronics, may be continuously rotated. (The frequency readout beam comes from a beam splitter up along the rotation axis and is directed over to the CH_4-stabilized laser. Electrical power comes to the rotating table through Hg-filled channels and a contactor pin assembly below the table.) The table rotation angle is sensed via 25 holes pierced in a metal band under the table. A single, separate hole provides absolute resynchronization each turn. The laser beat frequency is counted for 0.2 sec under minicomputer con-

Work of the U. S. Government
Not subject to U. S. copyright

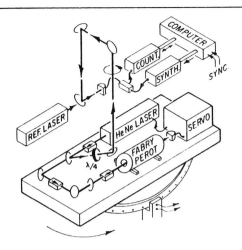

FIG. 1. Schematic of isotropy-of-space experiment. A He-Ne laser (3.39 μm) is servostabilized to a transmission fringe of an isolated and highly stable Fabry-Perot resonator, with provision being made to rotate this whole system. A small portion of the laser beam is diverted up along the table rotation axis to read out the cavity length via optical heterodyne with an "isolation laser" which is stabilized relative to a CH_4-stabilized reference laser. The beat frequency is shifted and counted under minicomputer control, these frequency measurements being synchronized and stored relative to the table's angular position. After 30 minutes of signal averaging the data are Fourier transformed and printed out, and the experiment is reinitialized.

trol after each synchronizing pulse, scaled and transferred to storage and display. A genuine spatial anisotropy would be manifest as a beat-frequency variation $\propto P_2(\cos\theta)$. The associated laser-frequency shift may be conveniently expressed as a vector amplitude at twice the table rotation frequency, f, of 1 per ~10 sec. Furthermore, its component in the plane perpendicular to Earth's spin axis should precess 360° in 12 h.

Our fundamental etalon of length is an interferometer which employs fused-silica mirrors "optically contacted" onto a low-expansion glass-ceramic[10] tube of 6-cm o.d. ×1-cm wall ×30.5-cm length. The choice of 50-cm mirror radii provides a well-isolated TEM_{∞} mode. Dielectric coatings at the mirrors' centers provide an interferometric efficiency of 25% and a fringe width ≈4.5 MHz. The interferometer mounts inside a massive, thermally isolated Al vacuum envelope. The environmental temperature is stable to 0.2 °C.

Fringe distortion due to optical feedback is prevented by a cascade of three yttrium-iron-garnet Faraday isolators, each having a return loss ≥26 dB. The laser is frequency modulated ≈2.5 MHz peak to peak at 45 kHz. Both first-harmonic and third-harmonic locking were tried, the unused one being a useful diagnostic for adjustment of the Faraday isolators. Based on the 200-μW available fringe signal, the frequency noise of the cavity-stabilized laser is expected (and observed) to be about 20 Hz for a 1-sec measurement, using a first-harmonic lock.

Our CH_4-stabilized "telescope-laser" frequency reference system achieves a comparable stability.[9] The random noise of the beat signal in a typical 20-min data block is observed to be ~3 Hz, compared with the laser frequency of almost 10^{14} Hz. To ensure absolute isolation of the cavity-stabilized and CH_4-stabilized lasers, the latter actually is used to phase lock a "local-oscillator" laser offset by 120 MHz. The ~35-MHz beat of this isolation laser with the cavity-stabilized laser is the measured quantity.

The *useful* sensitivity of our experiment is limited mainly by two factors: drift of the interferometer (~−50 Hz/sec) and a spurious nearly sinusoidal frequency shift at the table rotation rate f. This latter "sine-wave" signal was typically about 200 Hz peak to peak, and arises from a varying gravitational stretching of the interferometer, if the rotation axis is not perfectly vertical. The centrifugal stretching due to rotation is −10 kHz at $f = (1 \text{ turn})/(13 \text{ sec})$ and implies a compliance ~10 times that of the bulk spacer material.

We find that taking data in blocks of N table rotations ($N \simeq 8-50$) is helpful in minimizing the cross coupling of these noise sources into the interesting Fourier bin at 2 cycles per table revolution (actually at $2N$ cycles per N table revolutions). Typically 10–20 blocks of N revolutions were averaged together in the minicomputer before calculating the amplitude and phase of the signal at the second harmonic of the table rotation frequency. The average result is an amplitude of $\cos 2\theta$ of ≈17 Hz (2×10^{-13}) with an approximately constant phase in the laboratory frame. A number of such $\frac{1}{2}$-h averages spanning a 24-h period are illustrated in Fig. 2 as radius vectors from the origin to the open circles. The noise level of each such average was estimated by computing the noise at the nearby Fourier bins of $2N \pm 1$ cycles per N table revolutions. For a $\frac{1}{2}$-h average ($N = 10$, averaged 10 times) the typical noise amplitude was 2 Hz with a random phase.

To discriminate between this persistent spuri-

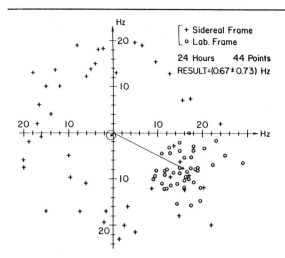

FIG. 2. Second Fourier amplitude from one day's data. The vector Fourier component at twice the table rotation rate is plotted as the radius vector from the origin to the open circles. After precessing these vectors by their appropriate sidereal angles they are plotted as the (+). For the 24-h block of data the average "ether drift" term is 0.67 ± 0.73 Hz, corresponding to $\Delta\nu/\nu = (0.76 \pm 0.83) \times 10^{-14}$.

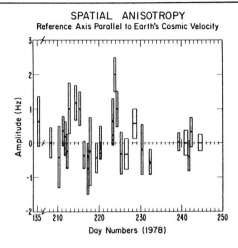

FIG. 3. Averaged data of isotropy-of-space experiment. Data such as those in Fig. 2, were averaged in blocks of 12 h (thinner bars) or 24 h (thicker bars). For completeness this figure includes data from diagnostic experiments before day 225. The data after day 238 represent near-ideal automatic operation of the present apparatus. A 1-Hz amplitude represents ~1.1 × 10^{-14} fractional frequency shift. The reference axis for the projection is the direction identified by Smoot et al. [11.0-h R.A. (right ascension), 6° dec. (declination)], Ref. 12.

ous signal (17-Hz amplitude at $2f$) and any genuine "ether" effect, we made measurements for 12 or 24 sidereal hours. We must rotate each vector to obtain its phase relative to a fixed sidereal axis prior to further averaging. Averaging after this rotation leads, as shown in Fig. 2, to a typical 1-day result below 1 ± 1 Hz. Averages for 24 h were sometimes quieter than 12-h averages, an effect which may be related to the observed 24-h period of the floor tilt ($\approx \mu$rad). A number of 12- and/or 24-h averages are shown in Fig. 3. These data include most of the points taken during various diagnostic experiments. The data taken after day 238 correspond to approximately "ideal" automated operation of the present apparatus. The lack of any significant signal or day dependence allows us to perform an overall average. This final result of our experiment is a null "ether drift" of 0.13 ± 0.22 Hz, which represents a fractional frequency shift of $(1.5 \pm 2.5) \times 10^{-15}$. From Eq. (1) we have $\Delta\nu(2\theta)/\nu = \frac{1}{2}[(g_2/g_1) - 1]$, so that our experimental result[11] can be written in the form $g_2/g_1 - 1 = (3 \pm 5) \times 10^{-15}$. We may conservatively use Earth's velocity around the sun to calculate the "expected" shift $\frac{1}{2}(v^2/c^2) \simeq \frac{1}{2}(10^{-8})$, which gives a null result[11] some 5×10^{-7} smaller than the classical prediction. This limit represents a 4000-fold improvement over the most sensitive previous experiment. This advance is due to smaller spurious signals in our experiment (2×10^{-13} instead of 10^{-9}), to superior data-processing techniques, and to superior long-term stability of the length etalon and reference laser.

The present sensitivity limit arises from two sources: the finite averaging time and some mechanical problems. To improve our result another decade by simple averaging would require 15 yr. The same decade improvement should be possible in several months' averaging, with improved mechanical design (rotation speed stable to 10^{-4} and the rotation axis actively stabilized to ±1″) and better vacuum stability inside the interferometer (to reduce the drift).

The recent discovery of a pure $P_1(\cos\theta)$ anisotropy in the cosmic blackbody radiation was interpreted as a Doppler shift produced by our motion (~400 km/sec) relative to a "privileged" inertial frame in which the blackbody radiation is isotropic.[12] If this velocity is considered to be the relevant one, our sensitivity[11] is ~3×10^{-9} and constitutes the most precise test yet of the Lorentz transformation. It will be especially interesting in the near future to develop techniques to look even more sensitively for some extremely small residual "preferred frame" or general-relativistic effects.[13]

We are grateful to James E. Faller for stimulation and for help in identifying our one-cycle-per-revolution spurious effect. One of us (J.L.H.) thanks J. Dreitlein, J. Castor, and R. Sinclair for useful discussions of general-relativistic effects; he is a staff member of the Quantum Physics Division of the U. S. National Bureau of Standard. The other (A.B.) acknowledges receipt of a NATO fellowship. We strongly thank P. L. Bender for his long-term interest in the experiment. The clever mechanical design work of C. E. Pelander has been indispensable. This research has been supported by the National Bureau of Standards under its program of precision measurement for possible application to basic standards, by the National Science Foundation, and by the Office of Naval Research.

(a) Permanent address: Laboratoire de l'Horloge Atomique, Orsay, France.

[1]A. A. Michelson and E. W. Morley, Am. J. Sci. 34, 333 (1887).

[2]G. Joos, Ann. Phys. 7, 385 (1930).

[3]For the works of H. A. Lorentz and G. F. FitzGerald, see, e.g., C. Møller, *The Theory of Relativity* (Clarendon Press, Oxford, 1972), 2nd ed., p. 27, and references therein.

[4]H. P. Robertson, Rev. Mod. Phys. 21, 378 (1949); H. P. Robertson and T. W. Noonan, *Relativity and Cosmology* (Saunders, Philadelphia, 1968).

[5]R. J. Kennedy and E. M. Thorndike, Phys. Rev. 42, 400 (1932).

[6]H. E. Ives and G. R. Stilwell, J. Opt. Soc. Am. 28, 215 (1938), and 31, 369 (1941).

[7]D. Newman, G. W. Ford, A. Rich, and E. Sweetman, Phys. Rev. Lett. 40, 1355 (1978).

[8]T. S. Jaseja, A. Javan, J. Murray, and C. H. Townes, Phys. Rev. 133, A1221 (1964).

[9]J. L. Hall, in *Fundamental and Applied Laser Physics: Proceedings of the 1971 Esfahan Symposium*, edited by M. S. Feld, A. Javan, and N. Kurnit (Wiley, New York, 1973), p. 463.

[10]CER-VIT is a registered trademark of Owens Illinois Inc., Toledo, Ohio.

[11]For comparison with earlier work we also omit a 0.43× sensitivity reduction factor associated with our 40° latitude and the assumed $P_2(\cos(\vec{v},\vec{1}))$ dependence.

[12]G. F. Smoot, M. V. Gorenstein, and R. A. Muller, Phys. Rev. Lett. 39, 898 (1977), and references therein.

[13]For example, Earth's gravitational field would provide a positive anisotropy signal due to the variation of the metric with height if the apparatus were to be rotated in a vertical plane, although it seems quite likely that this small fractional frequency shift $(GM/R)(1/c^2)(L/R) \simeq 3\times 10^{-17}$ with a $P_2(\cos\theta)$ signature would be toally obscured by the $P_1(\cos\theta)$ signal arising from gravitational stretching of the interferometer. We note that the present experiments concern the behavior of "rigid measuring rods" as distinguished from "atomic clocks" used, e.g., in R. F. C. Vessot and M. W. Levine, Center for Astrophysics (Cambridge, Massachusetts) Report No. 993, 1978 (to be published), and references therein, and in *Experimental Gravitation* (Academia Nazionale dei Lincei, Rome, 1977), p. 372.

AMERICAN JOURNAL of PHYSICS

VOLUME 39/10
OCTOBER 1971

October 1971

A Velocity of Light Measurement Using a Laser Beam

D. S. EDMONDS, JR.
R. V. SMITH
Department of Physics
Boston University
Boston, Massachusetts 02215

(*The apparatus described in this article won a Merit Award at the 1971 AAPT Apparatus Competition—Editor*)

INTRODUCTION

The velocity of light is important in a general physics course because it plays a fundamental role in electromagnetism and relativity and because its measurement forms a significant and dramatic chapter in the evolution of physics. However, the fact that light takes time to go from one place to another is sufficiently far outside the daily experience of most students that its reality is hard to get across. Commercially available systems employing modifications of the Michelson rotating-mirror method are not very useful for class demonstrations because the observations must be made by each student through an eyepiece and the darkening of the field at certain mirror speeds is unconvincing evidence for the finite speed of light. Furthermore, the student finds measuring the revolutions per minute of the mirror a bothersome complication.

We have adopted a system that, in effect, is Galileo's original experiment but employs electronic light chopping and detecting equipment instead of hooded lanterns and the human eye[1,2] (Fig. 1). For our setup, the time interval is measured on an oscilloscope display (Fig. 2). Tyler[1] had found this method of presentation dramatic and well suited for demonstrations before an audience. In addition, we use a laser light beam because of its high degree of collimation, which allows easy and rapid initial alignment of the optical system and leaves no doubt in the observer's mind about its path. Whether this path is back and forth along the building corridor or around a series of mirrors in the lecture hall, the student can see it and given a surveyor's tape, can measure it on the spot. On dividing this distance by the time interval as read from the oscilloscope display, he can obtain a value for the velocity of light within a few percent of 3×10^8 m/sec. The evidence for a high but finite velocity of propagation of light by this demonstration experiment is very convincing.

PRINCIPLE OF OPERATION

To measure its propagation velocity the laser beam is first "marked" or modulated by chopping

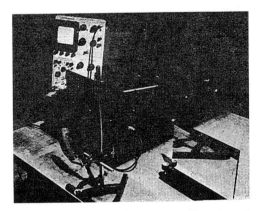

FIG. 1. General view of the apparatus. The mirror that reflects the beam for its second round trip across the laboratory is seen at the right.

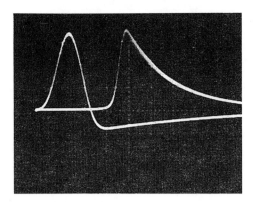

FIG. 2. Typical display with dual-trace oscilloscope. Horizontal sweep time is 0.2 μsec/cm.

it with a high-speed rotating mirror. The light reflected from this mirror travels a distance of 25.5 in. to an opaque screen, where a small hole transmits pulses of light as the beam sweeps over it. Since the mirror rotates at about 30 000 rpm and the beam diameter is of the order of 0.4 cm, the risetime of the light pulse (i.e., the time required for the beam to sweep over the edge of the hole) is less than 0.5 μsec. These light pulses strike a glass plate mounted at 45° to the beam direction so as to reflect a small fraction to a photodiode. The output of this "near" photodiode both triggers the oscilloscope and produces a pulse on the screen which marks the instant of departure of the light flash from the glass plate. Most of the light passes through the glass plate to travel over the measured path along the corridor, around the lecture hall, or in the laboratory room, as illustrated in Fig. 3. A path length of not less than 100 m should be provided in order to produce a good oscilloscope display. With a sweep speed of 0.2 μsec/cm, the leading edge of each pulse occupies about 1 cm on the screen. Since a path length of 100 m results in a delay of 0.33 μsec, the displacement between the initial and final pulses should be a little more than a centimeter. The final pulse is produced by a second photodiode on which the light falls after traveling the measured path. To obtain the display, these pulses are fed into the two channels of a dual-channel plug-in unit in the oscilloscope, which is set for the "alternate" mode. Thus one trigger pulse initiates a sweep showing the "near" photodiode output and the next pulse initiates the alternate sweep showing the output of the second or "far" photodiode. By holding a mirror directly in front of the apparatus the beam may be reflected to the "far" photodiode at once. Since the light path then has nearly the same length as the beam reflected to the "near" photodiode, the pulses fall on top of one another. In this way it is shown that the observed pulse separation is really due to the time required for light to traverse the measured path.

THE APPARATUS

The laser and rotating mirror assembly are mounted on a ½-in. thick aluminum base plate 18 in. wide by 30 in. long. The plate is supported on the bench by three leveling screws primarily to prevent rocking. A vertical screen of ⅛-in. aluminum sheet is mounted on the base plate 2 ft from the rotating mirror. Immediately the other side of it, a shelf (also made of ½-in. aluminum plate) supports the two electronic utility boxes (Bud AU-1029) containing the solid-state photocells, associated batteries, and circuitry shown in Fig. 4.

The Spectra-Physics Model 130B cw helium–neon gas laser used in the experiment has a rated power output of 0.75 mW at 6328 Å. Its beam at a height of $4\frac{5}{16}$ in. above the base plate is aimed at the center of the rotating mirror in a Leybold motor-mirror assembly.[3] The motor unit is held at its midsection by a bracket that is pivoted between trunions at the rear of the base plate. The bracket may be tilted by a $\frac{1}{4}$–40 adjusting screw so that the beam reflected from the rotating mirror can be aimed at the chopping hole in the aluminum screen. Only one reflective

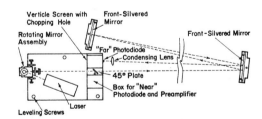

FIG. 3. Schematic diagram of the experimental arrangement showing a path length equal to four transits of the laboratory room.

surface of the mirror is used; the other is painted with a flat black paint. The motor's lower section hangs below the edge of the base plate and the benchtop.

Because the chopping hole in the aluminum screen of our setup has a diameter of $\frac{3}{16}$ in. it is so much larger than the beam diameter that the transmitted pulse has a lateral sweep. Therefore, great care is necessary in aligning that part of the sweep falling on the "far" photodiode so that it corresponds to the position of the beam being reflected by the glass plate into the "near" photodiode; otherwise a spurious delay appears between the "near" and "far" signals. This difficulty can be reduced by drilling the chopping hole no larger than the beam diameter. For the glass plate we use a Cenco No. 85540 index-of-refraction plate (7 cm$\times$7 cm$\times$6 mm) mounted on a piece of $\frac{1}{2}$-in. brass. The diodes are the Edgerton, Germeshausen, and Grier (EG&G) type SD100 solid state units, which have a 10-nsec risetime when reverse biased with 90 V.[4]

The photodiode circuits are shown in Fig. 4. The photodiodes are reverse biased with 90 V from portable "B" batteries to insure adequate response speeds. They are mounted inside the utility boxes with $\frac{3}{8}$-in. rubber grommets not only to insulate them electrically but also to isolate them from external vibrations.

The amount of light reflected to the "near" photodiode by the glass plate is insufficient to produce an output pulse capable of triggering the oscilloscope, even with the largest permissible load resistor; therefore, a small pulse amplifier with its batteries and a switch is incorporated with the 90-V bias battery in the utility box for the "near" photodiode. The "far" photodiode box contains only the bias battery, bypass capacitor, and load resistor. This latter's value (2200 Ω) is determined to be that for which the risetime of the diode circuit just equals the risetime inherent in the light-chopping system. The triggering is external so that the trace will be initiated by the "near" diode signal regardless of whether this signal or that from the "far" diode is being displayed.[5]

Two 8$\times$10 in. front-silvered mirrors[6] mounted in brass holders, each supported by three $\frac{1}{4}$-40 adjusting screws, served to change the direction of the beam path. The laser beam diverges suffi-

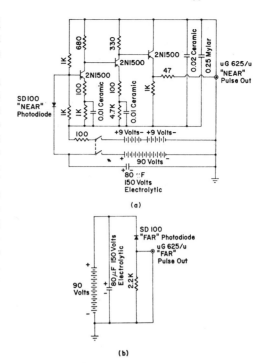

Fig. 4. The photodiode circuits. (a) "Near" photodiode and associated pulse preamplifier. (b) "Far" photodiode.

ciently over a path length of 200 m so that a condensing lens of approximately 20-cm focal length is needed to concentrate the light on the photosensitive surface of the "far" photodiode. The oscilloscope should be fast enough to handle pulses with risetimes of a hundred nanoseconds. Also it should have slave sweep and if possible, dual trace capabilities.[7]

OPERATION

Initially with the power off, the rotating mirror is turned by means of a key until the reflected beam is directed at the chopping hole. The motor mount and glass plate are then adjusted until the beam passes directly through the hole and a portion is reflected from the glass plate's front surface into the "near" photodiode. The beam will now be faintly visible in a darkened room, and the large fixed mirrors may be placed so as to direct it along its path and return it to the "far" photodiode. The preliminary alignment is completed by focusing the return beam through a condensing lens onto the sensitive surface of this diode.

The mirror motor, oscilloscope, and pulse preamplifier are turned on and the apparatus is further adjusted to give a satisfactory "near" pulse. The position of the glass plate also may need slight readjustment. The pulse from the "far" photodiode should now be detectable. Finally, the large fixed mirrors are readjusted to obtain minimum spacing between pulses.

A typical trace is shown in Fig. 2. The oscilloscope sweep speed is 0.2 μsec/cm. Since each large graticule division is 1 cm, the displacement of the pulses is 0.64 ± 0.01 μsec between their starting points. Note that because of the pulse preamplifier in one channel, the leading edges do not have the same shape. But their starting points may be shown to coincide by reducing the path difference to zero with an auxiliary mirror as described earlier. The path length for the time interval recorded in Fig. 2 was 189.3 ± 0.1 m. From these data we obtain $2.96\pm0.05\times10^8$ m/sec for the speed of light. The precision of this result is clearly limited by the accuracy with which the time interval can be read from the oscilloscope pattern, but it appears good enough for lecture demonstrations or undergraduate laboratory measurements.

ACKNOWLEDGMENTS

We would like to thank Mr. R. Allan O'Neill of the Boston University Physics Department Shop for his help in the construction of this apparatus and Dr. P. Bruce Newell of Edgerton, Germeshausen, and Grier for making the SD-100 photodiodes available to us.

[1] C. E. Tyler, Amer. J. Phys. **37**, 1154 (1969).

[2] R. V. Smith and D. S. Edmonds, Jr., Amer. J. Phys. **38**, 1481 (1970).

[3] Available in this country through the Lapine Scientific Company as catalog No. Y-47641 (price $225.00).

[4] These diodes are an obsolete experimental type obtained from EG&G without charge. The equivalent currently available type is the SGD-100, which is identical to the SD-100 except for the addition of a guard ring around the active cathode area and a somewhat better performance. The price is $40 each.

[5] If a dual-trace plug in is not available, the "near" pulse may be used to trigger the oscilloscope and only the "far" pulse displayed on the screen. Since the oscilloscope sweep starts immediately upon application of the trigger, the travel time is given by the displacement of the "far" pulse from the start of the trace, but the resulting display is less convincing.

[6] Available from the Edmund Scientific Co. (catalog No. 40,067) at $6.00 each.

[7] We used Tektronix models 585, 581, and 543. The 85-mc bandwidth of the 580 series is more than adequate; the 33-mc response of the model 543 is satisfactory. For each of these models we used the 1A1 dual-trace plug in.

Another velocity of light experiment

B. G. Eaton, Philip A. Johnson, and Noel J. Petit
School of Physics and Astronomy, University of Minnesota, Minneapolis, MN 55455

Many of the lecturers in our elementary physics courses like to use the time signal from WWV when talking about mass, length and time in the beginning of their courses. The old vacuum-tube receiver used for this purpose was difficult to tune and often blamed for poor performance. It was subsequently replaced by a Timekube[1] from Radio Shack. The Timekube is a crystal-controlled, solid-state receiver capable of receiving on 5, 10, and 15 mHz. By connecting a 0.1 μF capacitor to the ungrounded terminal of the speaker coil, one can bring a signal out through a BNC connector so it is available for observation on an oscilloscope.

When the new system was tried in lecture for the first time, however, our great expectations quickly evaporated. Upon further investigation, it was found that the Timekube's antenna was inadequate for our needs. We then purchased a CB, 0.64 wave, ground-plane antenna from Radio Shack[2] and mounted it on the roof of the physics building in an unobstructed area. The combination of Timekube and antenna was very successful, as can be seen in Fig. 1. The second pulse is not a reflection of the first. This can be verified by counting the number of cycles in each pulse: the first pulse is five cycles of a one-kHz signal originating at WWV in Fort Collins, CO; the second pulse is six cycles of a 1.2-kHz signal which originates at WWVH in Kauai, HI. Since both pulses are emitted simultaneously, the difference in arrival times can be used to calculate the speed of the radio waves.

If the earth were flat, the distance calculations would be easy, but the earth is round and the radio waves bounce back and forth between its surface and the ionosphere.[3] The distance calculation starts with a great circle calculation using:

$$S = 60 \cos^{-1}(\cos L_a \cos L_b \cos P \pm \sin L_a \sin L_b)$$

where S is the great circle distance in nautical miles, L_a and L_b are the latitudes of points a and b, and $P = L_{oa} - L_{ob}$ is the difference in longitude of points a and b. The + sign is used if a and b are on the same side of the equator, and the − sign is used if they are on opposite sides of the equator. In order to make this calculation you need to know the longitude and latitude of your receiver and of the two transmitters. The data for Fort Collins is: latitude, 40° 40′ 49″ N and longitude, 105° 02′ 27″ W. Kauai's latitude and longitude are 21° 59′ 26″ N and 159° 46′ 00″ W, respectively. The resulting great circle distance can be converted from nautical miles to kilometers by multiplying by 1.8522.

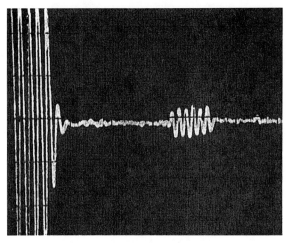

Fig. 1. Oscilloscope picture of trace showing the "tics" from WWV and WWVH. The sweep rate is 5 msec/cm.

TABLE I		
Speed-of-light calculations for various bounce altitudes		
Altitude	c m/sec	Percent error
Great Circle only	2.84×10^8	−5.4
110 km	2.91×10^8	−3.2
200 km	3.04×10^8	+1.2
250 km	3.14×10^8	+4.5

The altitude of the layer that will reflect the radio waves depends upon the day, the hour, and location on the globe.[4] In general, the layer will be between 100 and 250 km high and there will be roughly one bounce per 1000 km of global distance. From the great-circle calculations one can find the nearest number of bounces and then use the Pythagorean theorem to find the actual path length. The actual difference in path length divided by the difference in arrival times will yield the wave velocity. The sweep rate in Figure 1 was 5 ms/cm, which yields 1.9×10^{-2} sec as the difference in travel times for the two pulses. Table I shows the calculated velocity of light for various bounce altitudes. One can see from these results that a good value of the bounce altitude is needed in order to obtain a reasonable value for the speed of light. This is actually a better experiment for determining the bounce altitude, but if one uses a value of 200 km then c can be found to within several percent. Conceptually this is not a difficult experiment to explain to students.

References

1. Trade name for a 5, 10, 15 MHz receiver for WWV signals sold by Radio Shack, #12-159, at $34.95.
2. Archer 0.64 - Wave, Ground Plane Antenna, Radio Shack, catalog #21-964, at $46.95.
3. A very useful pamphlet on the WWV system and how to make these calculations is available from the Government Printing Office. SD Cat. No. C13, Technical Note 668, Project 2770128, May 1975, Price $1.05. Sup. of Doc., U.S. Government Printing Office, Washington, D.C. 20402.
4. Data on heights of bounce layers by day, time, and location is available from: World Data Center, NOAA, Dept. of Commerce, Boulder, CO 80302.

Reprinted with permission from *American Journal of Physics* 51, 1003–1008, ©1983 American Association of Physics Teachers.

Measuring the speed of light by independent frequency and wavelength determination

Harry E. Bates
Department of Physics, Towson State University, Baltimore, Maryland 21204

(Received 9 August 1982; accepted for publication 7 January 1983)

The most recent measurements of the speed of light have been based on a separate determination of the frequency and wavelength of laser radiation. As a result of the increased precision and the imbalance in precision between frequency and wavelength measurements, the speed of light in a vacuum c (299,792,458 m/s) has been proposed as a defined standard constant connecting time and length measurements. This paper reports on a student experiment at the intermediate undergraduate level that contains the fundamental elements found in the most recent determination of c. Harmonic mixing is utilized for the frequency determination and interferometry for the wavelength determination of a 9-GHz electromagnetic source.

I. INTRODUCTION

The measurement of a fundamental constant of nature is certainly an appropriate laboratory experiment for the intermediate undergraduate physics student. One of the most interesting of the fundamental constants is the speed of light. It is the bridge between space and time in relativity; it is involved in the conversion between electrostatic and electromagnetic units; it relates mass to total particle energies and its precise determination plays a central role in tests of quantum electrodynamics. Within the last decade successes in independent wavelength and frequency measurements in the optical region have led to the most precise determination of the speed of light.

Numerous measurements of c have been published during the development of modern physics.[1] Techniques for its measurement have been diverse and have included observations in astronomy[2] as well as interferometry,[3] spectroscopy,[4] and more straightforward distance plus propagation time measurements.[5] In addition, a number of articles have appeared reporting on student measurements.[6-8] One of the most significant results of the invention of the laser has been its application in techniques for measuring c. These recent experiments have yielded precisions of the order of 100 times better than before.

The purpose of this paper is to report on an experiment developed at Towson State University for the measurement of the speed of light in an intermediate undergraduate physics laboratory. Although a number of other experiments for measuring c at this level have been reported, this variation contains significant elements used in the most recent laser determination. That is, an interferometric measurement of wavelength is made. At the same time, an independent measurement of frequency is made using a harmonic mixing technique.

II. LASER TECHNIQUE

The most recent advances in the technique for measuring the speed of light have been made by Evenson and his co-workers at the National Bureau of Standards Laboratories in Boulder, Colorado.[9,10] They were made possible by

Table I. Laser measurements of the speed of light.[a]

Measurement[b]	Laser	Wavelength (vacuum) nm	Ref.	Frequency THz	Ref.	Velocity of light m/s
NBS, Boulder Col (Kr Center of Gravity)	He–Ne	3392.231376(12)[c]	12, 9	88.376181627(50)	9, 13	299792456.2(1.1)
NBS, Boulder Col (Kr Peak)	He–Ne	3392.231404(12)	12, 9	88.376181627(50)	9, 13	299792458.7(1.1)
NPL	CO_2 R(12)	9317.246340(6)	14	32.176079482(27)	15	299792458.8(0.2)
MIT, NBS	CO_2 R(14)	9305.385613(70)	16	32.217091275(24)	13, 17	299792457.6(2.2)
NRC	He–Ne	3392.231400(20)	18	88.376181570(200)	19	299792458.1(1.9)

[a] Data for this table obtained from Evenson, Ref. 10.
[b] Location of measurement: NBS, Boulder, Colorado is the National Bureau of Standards; NPL is the National Physical Laboratory at Teddington, England; and NRC is the National Research Council at Ottawa, Canada.
[c] Uncertainties in numerical data are shown in parentheses.

several developing technologies. First, the science of optical interferometry and the definition of the standard meter in terms of the wavelength of a transition in atomic krypton-86 made it possible to measure lengths to a few parts in 10^9. Next, as a result of intense investigations of microwaves with application to radar systems during World War II, methods of frequency measurement were extended to the microwave region using harmonic generation and difference frequency mixing techniques. Advances in radar technology also led to the development of the maser and the laser. With the invention of stabilized lasers, oscillators having frequencies in the tens to hundreds of terahertz became available. A key element in extending the measurement of frequencies into the optical range was the development of suitable metal–insulator–metal (MIM) point contact mixer diodes first used at laser frequencies by Javan.[11] By using a chain of harmonic mixers including (MIM's) and other nonlinear devices, Evenson and his co-workers were able to measure these extremely high-frequency laser sources by referencing them to lower frequency standard cesuim clocks. Thus they were able to carry out the first independent frequency and wavelength measurements on electromagnetic radiation in the near infrared. These measurements when multiplied together yielded the speed of light with a significant improvement in precision. The primary limitation in the precision of this measurement was the uncertainty in wavelength due to a slight asymmetry of the krypton line which was used as the length standard.

Since this first direct frequency measurement of an optical source, a number of other measurements have been made. Table I summarizes recent frequency and wavelength values obtained as well as values computed for the speed of light. The variation in the computed value of c due to the asymmetry of the krypton line is illustrated in the first two lines of the table.

It is clear that the fundamental experimental physics in all of this can be reduced to two kinds of measurements: (1) interferometric measurement of wavelength and (2) harmonic frequency mixing including the counting of a standard oscillator and observation of zero beats between harmonics of this oscillator and the primary electromagnetic source. The design of a student experiment was accomplished with these two elements in mind.

III. THEORY

The magnitude of the phase velocity of a wave can be expressed in terms of the distance of travel in one cycle (the wavelength λ) divided by the time of one cycle (the period T)

$$v = \lambda / T. \qquad (1)$$

Since the frequency is just the reciprocal of the period equation (1) can be rewritten as

$$v = \lambda f. \qquad (2)$$

Thus independant measurements of wavelength and frequency yield the phase speed of a wave.

A Michaelson interferometer can be utilized to determine the wavelength. It is desirable to use wavelengths as short as possible so as to count as many fringes as possible in each single interferometer run. This enhances the precision of the wavelength measurement and leads to the use of very high frequencies in state-of-the-art measurements. Such frequencies in the microwave region of the electromagnetic spectrum and beyond are too high for direct

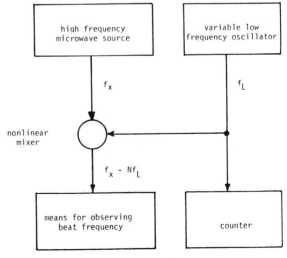

Fig. 1. System for measuring high frequencies using nonlinear harmonic mixing.

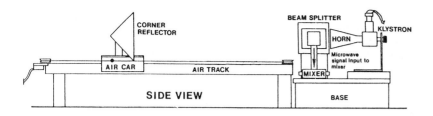
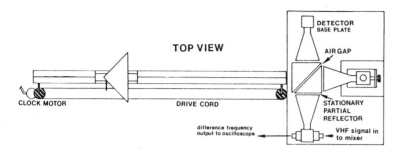

Fig. 2. Microwave Michaelson interferometer with detector and nonlinear mixer for frequency measurement.

counting methods using electronic gates; so an indirect technique is required.

A very high frequency can be measured by the observation of nonlinear beat phenomena in a mixer. Consider the diagram of Fig. 1. A high-frequency microwave source produces a signal of frequency f_x to be measured. This signal is mixed with the output of a variable low-frequency oscillator. The oscillator produces a signal whose frequency is low enough to be measured using a frequency counter. The mixer, through its nonlinear action, produces harmonics of the input signals as well as sum and difference frequencies between the input signals and/or their various harmonics. One sequence of such signals of special interest has the form $f_x - Nf_L$, where N is the harmonic number, ($N = 1,2,3,...$).

If Nf_L is very close to f_x, the difference $f_x - Nf_L$ produces a low-frequency beat signal that can be observed on an oscilloscope or other suitable detector. The observation of this beat signal together with a measurement of f_L for a sequence of f_L's provides a means for indirectly determining f_x. For example, suppose two low frequencies f_1 and f_2 were observed to produce successive beats with some unknown f_x and with $f_2 > f_1$ such that

$$f_x = Nf_2 \tag{3}$$

and

$$f_x = (N+1)f_1, \tag{4}$$

where N is the harmonic number associated with the higher of the low frequencies f_2. Eliminating f_x in (3) and (4) we have

$$Nf_2 = (N+1)f_1 \tag{5}$$

or

$$N = f_1/(f_2 - f_1). \tag{6}$$

Thus by determining f_1 and f_2, N can be found. This should be rounded to the nearest whole number and then put into (3) to determine f_x.

With a good technique in hand for measuring very high frequencies, all that is required for completing the measurement is a microwave Michaelson interferometer. A detailed description of the system that we have used is given next along with a discussion of specific instruments used for the frequency measurements.

IV. PROCEDURE AND APPARATUS

The apparatus for this experiment is divided into two major sections, the wavelength measuring interferometer and the frequency measuring instrumentation. I will discuss each separately and then explain the procedure to be used in a measurement.

The wavelength is measured using a microwave Michaelson interferometer very similar to a Doppler velocity measuring system previously reported by the author.[20] A diagram is shown in Fig. 2. A Raytheon Heathkit klystron model no. RK2K25 part of a Heath model EPW-25 microwave generator powered by a Heath ID-17 supply is used to generate an electromagnetic wave at a nominal frequency of 9 GHz. The beam from the horn attached to the klystron is directed to a paraffin wax beam splitter (see Fig. 3) so that about one-half the intensity is reflected to the other surface

Fig. 3. Beam splitter, klystron source on the left, mixer and detector horn farside of beam splitter.

Fig. 4. Interferometer in operation—note corner reflector on air car driven by cord and pulley system.

Fig. 5. Block diagram of frequency measuring system.

of the wax prism which serves as a stationary partial reflector. I originally used a flat screen for this reflector (shown in the figure) but discovered that Fresnel reflection from the wax was suitable. The other half of the beam is transmitted by the wax prisms to a movable corner reflector. The corner reflector is mounted on a 1-m (or longer) air track by means of a standard air track car. A photograph of the complete interferometer assembly being operated by two students is shown in Fig. 4. Note the motor-driven chopper wheel in front of the transmitting klystron horn. This is not needed in the basic setup but represents one kind of improvement possible in the system. A discussion of these possible improvements is given in the next section. The reflected beams from both the movable and fixed reflectors are recombined in the detector via the beam splitter in the normal Michelson arrangement except no compensator plate is needed. Fringes are detected and recorded. In a typical measuring run, the air car is allowed to move over a measured distance and the number of fringes are counted from the chart recording.

The ultimate precision of the measurement is determined by the student's care and technique used in this wavelength measurement because the frequency measurement is fundamentally more precise.

Frequency is measured using a harmonic mixing scheme. A block diagram of the apparatus is shown in Fig. 5. A portion of the microwave signal transmitted through the wax prism partial stationary reflector of the interferometer is collected using a receiving horn and directed via waveguide to a Hewlett–Packard 934A harmonic mixer. A signal at about 1 GHz is generated using a Hewlett–Packard 8614B variable frequency oscillator and also directed to the mixer. The mixer contains a nonlinear mixing crystal which generates harmonics of the input signals as well as sum and difference frequency components. The output of the mixer is fed to the vertical input of an oscilloscope which is then used to detect the presence of a zero beat pattern. This pattern is only present when the frequency of one of the harmonics of the oscillator is very close to the microwave frequency.

V. POSSIBILITIES FOR IMPROVEMENT IN THE APPARATUS

One very attractive feature of this experiment is that students can take the responsibility of working on improving various aspects of the measurement and thereby learn more advanced techniques. The proof of their success is self-evident in the increase in precision of their measurement contrasted to measurements made with the basic setup. Some improvements that have been attempted or suggested include:

(1) Improvement in resolution and signal-to-noise ratio of the microwave fringes by modulating the klystron (see the chopper in Fig. 4) and using a lock-in amplifier for fringe detection.

(2) Improvement in the wavelength measurement by coupling an optical interferometer to the movable arm of the microwave interferometer referencing the distance measurement to the wavelength of a known laser line.

Table II. Typical experimental data collected.

Run No.	No. of fringes	2X Reflector travel (cm)	Zero beat freq. (MHz) f_1	Zero beat freq. (MHz) f_2	Harmonic No.
1	25	86.35		863.59	10
2	28	97.61	958.91		9
3	30	104.79		863.66	10
4	24	83.66	958.64		9
5	30	105.09		863.73	10
6	19	66.69	958.58		9

Table III. Typical calculated values based on measurements.

Run No.	Microwave wavelength (cm)	Microwave frequency (MHz)	Velocity of light cm/s (10^{10})	Percent error
1	3.454	8635.9	2.983	−0.5
2	3.486	8630.2	3.008	0.34
3	3.493	8636.6	3.017	0.64
4	3.486	8627.8	3.008	0.34
5	3.503	8637.3	3.026	0.94
6	3.510	8627.3	3.028	1.00

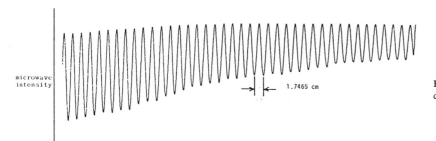

Fig. 6. Sample of interferometer fringes recorded during run No. 3.

(3) Improvement of wavelength determination by increasing the total path difference change for wavelength measurements.

(4) Improvement in the frequency measurement by referencing the counter gate time to a standard time signal.

(5) Phase locking the klystron to improve frequency stability.

(6) Investigation of near-field effects by comparing measurements made at a considerable distance from the klystron to those nearby. These experiments are made possible with the increased sensitivity afforded by the use of lock-in detection.

VI. EXPERIMENTAL RESULTS OF SEVERAL STUDENT RUNS

Many determinations of the speed of light have been made by students in the past 2 years. Most used a length reference which consisted of the meter scale mounted on the air track. A sample of data is given in Table II to illustrate the degree of precision that is routinely obtained. Calculated values based on these data are given in Table III. Figure 6 shows a record of the interferometer fringes and Fig. 7 shows a typical zero beat pattern observed when making frequency measurements.

Four to six meaurements were made at a single harmonic number of the local oscillator during each experimental run. The estimated limit of precision of the wavelength

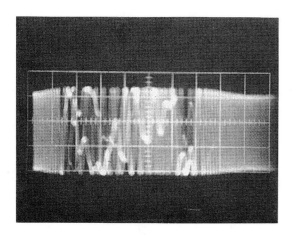

Fig. 7. Typical zero beat pattern observed. Oscilloscope is triggered on the power line because 60-Hz stray fields are the primary source of fm noise in the klystron.

measurement was ± 0.002 cm and that of the frequency measurement ± 0.5 MHz. However, during the afternoon when data were collected the standard deviation of the six wavelength measurements was about 0.02 cm and the standard deviation of the six frequency averages was about 5 MHz. This indicates that some random experimental error created nearly an order of magnitude more uncertainty than can be accounted for on the basis of the estimated limit of precision. The average value of the speed of light computed from these data was 3.01×10^{10} cm/s with a standard deviation of 1.6×10^8 cm/s. This mean value is 1.2×10^8 cm/s higher than the accepted value. Thus the experiment yielded a result within one standard deviation of the accepted value. Furthermore, the results clearly indicated that the precision in the wavelength determination was the dominant source of error in the experiment.

VII. CONCLUSIONS

The experiment provides the student the opportunity to measure the velocity of light using the same fundamental techniques applied in most recent years. The precision is only about a part in 10^2 or 10^3 depending on the care with which the wavelength measurement is made. This is six or seven orders of magnitude worse than the best measurement in terms of the standard meter. But the character of the measurement is the same. The same imbalance in fundamental precision exists in this student measurement as in the best measurement. That is, given the level of sophistication of the experimental apparatus, the frequency measurements are much more precise than wavelength measurements. This then leads the student to the same question now facing the scientific and technical community, should the speed of light be adopted as a defined constant forcing length measurements to be expressed in terms of time.

ACKNOWLEDGMENTS

The author would like to thank Kenneth M. Evenson of the National Bureau of Standards for many helpful conversations and correspondence. He would also like to thank the Towson State University Faculty Research Committee for its support. In addition he would like to thank Thomas O. Krause for reading drafts of the manuscript and making many helpful suggestions. Last, but not least, he would like to thank Thomas K. Hemmick and Donald L. Fisher, two students that made significant improvements to the apparatus during the course of this work and who agreed to be photographed while their intermediate laboratory experiment was in progress.

[1] J. F. Mulligan, Am. J. Phys. **44**, 960 (1976).
[2] M. Born and E. Wolf, *Principles of Optics* (Pergamon, New York, 1959), p. 11.
[3] K. D. Froome, Proc. R. Soc. London Ser. A **247**, 109 (1958).
[4] E. K. Plyler, L. R. Blaine, and W. S. Connor, J. Opt. Soc. Am. **45**, 102 (1955).
[5] E. Bergstrand, Nature **165**, 405 (1950).
[6] L. Bergel and S. Arnold, Am. J. Phys. **44**, 546 (1976).
[7] D. N. Page and C. D. Geilker, Am. J. Phys. **40**, 86 (1972).
[8] W. F. Huang, Am. J. Phys. **38**, 1159 (1970).
[9] K. M. Evenson, J. S. Wells, F. R. Petersen, B. L. Danielson, G. W. Day, R. L. Barger, and J. L. Hall, Phys. Rev. Lett. **29**, 1346 (1972).
[10] K. M. Evenson, private communication to be published in Proceedings of NATO School of Fundamental Constants held in Erice, Sicily, Nov. 1981.
[11] V. Daneu, D. Sokoloff, A. Sanchez, and A. Javan, Appl. Phys. Lett. **15**, 398 (1969).
[12] R. L. Barger and J. L. Hall, Appl. Phys. Lett. **22**, 196 (1973).
[13] K. M. Evenson, J. S. Wells, F. R. Petersen, B L. Danielson, and G. W. Day, Appl. Phys. Lett. **22**, 192 (1973).
[14] P. T. Woods, B. W. Jolliffe, W. R. C. Rowley, K. C. Shotton, and A. J. Wallard, Appl. Opt. **17**, 1052 (1978).
[15] T. G. Blaney, C. C. Bradey, G. J. Edwards, B. W. Jolliffe, D. J. E. Knight, W. R. C. Rowley, K. C. Shotton, and P. T. Woods, Nature **251**, 46 (1974).
[16] J. P. Monchalin, M. J. Kelly, J. E. Thomas, N. A. Kurnit, A. Szoke, and A. Javan, Opt. Lett. **1**, 5 (1977).
[17] F. R. Petersen, D. G. McDonald, J. D. Cupp, and B. L. Danielson, *Laser Spectroscopy*, Proceedings of the Vail, Colorado Conference, edited by R. G. Brewer and A. Mooradian (Plenum, New York, 1973).
[18] K. M. Baird, D. S. Smith, and W. E. Berger, Opt. Commun. **7**, 107 (1973).
[19] K. M. Baird, D. S. Smith, and W. E. Berger, Opt. Commun. **31**, 367 (1979).
[20] H. E. Bates, Am. J. Phys. **45**, 711 (1977).

Reprinted with permission from *American Journal of Physics* 55, 853-854, ©1987 American Association of Physics Teachers.

A pulser circuit for measuring the speed of light

M. E. Ciholas and P. M. Wilt
Centre College, Danville, Kentucky 40422

The speed of light c is unquestionably one of the most important physical constants. Its measurement can be effectively included in an undergraduate physics curriculum using several different methods. One such method employs a pulsed light-emitting diode (LED) and oscilloscope with ns/cm sweep speeds to measure the time required for a light pulse to travel a few meters.

Briefly, an electronic pulser circuit produces an electrical pulse of approximately 20 ns duration. This pulse is split into two parts. One part triggers the horizontal sweep of a fast oscilloscope while the other generates a light pulse in a LED. This light pulse is collimated and travels a few meters through a retroreflector to a focusing lens and photomultiplier detector. The detector output is fed to the verticle input of the oscilloscope where a pulse is thus produced on the screen. The position of this pulse is noted and the retroreflector moved a measured distance d. The pulse position shifts an amount Δt relative to its previous position. This time Δt is measured using the oscilloscope time base; c is then obtained by dividing $2d$ by Δt because moving the retroreflector a distance d changes the light path by $2d$. Note that the time difference Δt is caused only by the change in light path d.

Tyler[1] describes in some detail how one can construct the electronic and optical components of such an apparatus to measure c in an undergraduate laboratory. Tyler's electrical approach was to use a multivibrator and avalanching-transistor circuit to pulse a LED. We have designed a new digital pulser circuit which, in our experience, offers increased stability and narrower pulse widths. This circuit is inexpensive, easy to build, and is described below.

In Fig. 1, invertors A and B, along with resistors R1 and R2, capacitor C1, and transistor Q1 form a square-wave oscillator whose frequency is determined primarily by the values of C1 and R1. To understand how the oscillator works, assume C1 is discharged. The characteristics of the

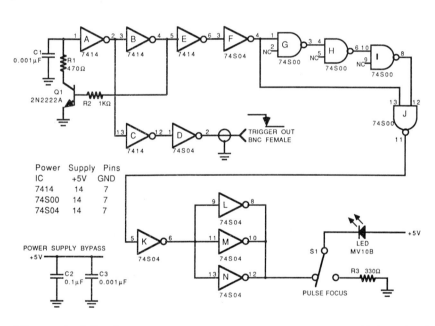

Fig. 1. Circuit diagram for pulsing a LED. The operation of this circuit is explained in text.

7414 IC cause pin 1 to float high when it is not driven low. C1 will thus be charged by pin 1 until C1 reaches a sufficiently high value to trigger inverter A. Pin 2 of A will then go low, and pin 4 of B will go high. A high state of pin 4 of B will cause Q1 to conduct and then discharge C1 through R1. After the voltage on C1 falls to a sufficiently low value, the output of A will go high and pin 4 of B will go low and Q1 will no longer conduct. C1 then begins to recharge and the cycle repeats; the oscillator frequency is about 250 kHz.

The output of A is also the source of the oscilloscope trigger signal. C and D of Fig. 1 twice invert and buffer the A output to isolate the trigger cable capacitance from the oscillator. The D output signal triggers the horizontal sweep of the oscilloscope.

We now return to the production of the LED pulser signal. Here E and F twice invert and buffer the output of B. The 74S04 IC has a much faster rise time than the 7414; this property sharpens the signal. The 74S00 IC is a quad nand gate each section of which has a propagation delay of about 5 ns. For convenience, the truth table for a nand gate is given in Table I. Note that the output is low only when both inputs are high.

At the output of F the signal splits: one part is fed immediately to pin 13 of J while the other travels through G, H, and I before reaching J. Pins 2, 5, and 9 of G, H, and I, respectively, are unconnected and assume the high state. Using Table I, one can see that if pin 4 of F goes high, pin 12 of J will become low about 15 ns later. If pin 4 of F goes low, pin 12 of J becomes high about 15 ns later. Thus the combination of G, H, and I invert the output of F and delay the signal by about 15 ns.

Let us suppose the output of F is low and then quickly goes high. At this instant pin 13 of J will go high and pin 12 of J will be high because its condition is determined by the output of F 15 ns earlier. Thus pin 11 of J will go low. After 15 ns the high state of pin 4 of F will propagate through G, H, and I to cause pin 12 of J to go low, and 11 of J to then go high. Thus the output of J goes low only for about 15 ns. The corresponding transition when the output of F goes from high to low does *not* cause the output of J to go low.

The pulsed-low output of J is inverted by K and split into three parts for current amplication by L, M, and N, which also reinvert the signal. The result is that when switch S1 is in the pulse position, a potential difference (approaching 5 V) of 15-ns duration across the LED produces a large current pulse. This current pulse generates a light pulse output from the LED. Many LED's designed to operate at a maximum continuous current of a few mA, can withstand much higher peak currents for nanosecond durations. A second switch position, focus, is provided to ground the LED through the 330-Ω current-limiting resistor R3. This setting produces a much brighter, continuous light output useful for visually focusing the optical components.

Table I. Truth table for nand gate.

Input 1	Input 2	Output
0	0	1
0	1	1
1	0	1
1	1	0

As in the construction of any high-frequency circuit, it is important to minimize lead lengths and stray capacitance. Our circuit utilizes point-to-point soldered wiring and is constructed on a prototype board of about 2-sq in. area.

Oscilloscope examination of the pulse shows that it is indeed quite sharp and exists for 16 ns. We have tested numerous LED's and find that the best combination of intensity, collimation, and pulse response is provided by the MV10B unit available from Arrow Electronics, Inc., 900 Brand Hollow Road, Farmingdale, NY. We used a Tektronix Model 2465 oscilloscope and an all-purpose Hamamatsu 931A photomultiplier tube. Somewhat stronger signals can be obtained with the more expensive, red-sensitive Hamamatsu R928 tube. The combination of electrical pulse width, PMT response time, and LED characteristics combine to give an observed light signal whose full duration is 30 ns.

Human judgment in the location of time center of the pulses appears to be the major factor limiting the accuracy of this method. Obviously a sharper light pulse signal would reduce errors in measuring Δt. We attempted to sharpen the light pulse signal by eliminating gates H and I to produce a 5 ns electrical pulse. This approach was unsuccessful because the output of J was not fully developed and the LED was not triggered. The use of 74AS00 and 74AS04 IC's, which have shorter propagation times, may produce a sharper pulse but we have not yet tried them.

Figure 1 also shows the power-supply connections for the various IC's. We used a simple + 5-V power supply and bypass. Jameco Electronics, 1355 Shoreway Road, Belmont, CA sells several suitable supplies for under $10.

We have used this method and the circuit described above in our optics and modern physics laboratories. The results are excellent. The circuit itself is very stable; we find it to be much more trouble-free than the avalanching-transistor approach. As an example of the accuracy of the approach, a recent sophomore-level optics class produced eight independent measurements of c whose average and standard deviation were $(3.00 \pm 0.03) \times 10^8$ m/s.

[1]C. E. Tyler, Am. J. Phys. **37**, 1154 (1969).